Rover 1S/60 gas turbine engine

Contributors:

Tom Barton	Doug Llewellyn
Ces Bedford	Tony Martin
Jack Belfit	Sid O'Neill
Warwick Bloor	Noel Penny
Chris Bramley	Phil Phillips
Peter Candy	Lofty Poole
Ron Hill	Joe Poole
Sid Hill	Don Proctor
Fred Jones	Bernard Smith
Spen King	Tony Worster

Cover: 1965 Rover-BRM sports car

Title page: The Rover Gas Turbines badge (BMIHT)

ROLLS-ROYCE HERITAGE TRUST

PISTONS TO BLADES
Small gas turbine developments by the Rover Company

Mark C S Barnard

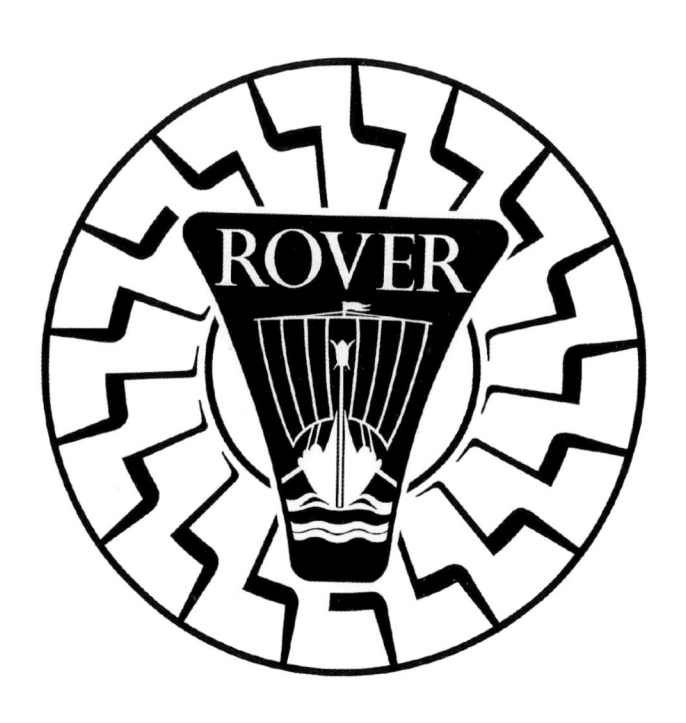

HISTORICAL SERIES No 34

Published in 2003 by the
Rolls-Royce Heritage Trust
PO Box 31 Derby England DE24 8BJ

© M C S Barnard
© Sandy Skinner, Appendix F
© Andrew Nahum, Appendix G

This book, or any part thereof,
must not be reproduced in any form without the
written permission of the publishers.

ISBN 1 872922 23 6

The Historical Series is published as a joint initiative by the
Rolls-Royce Heritage Trust and the Sir Henry Royce Memorial Foundation.

Previous volumes published in the Series are listed at the rear, together with
volumes available in the Rolls-Royce Heritage Trust Technical Series.

Books are available from:
Rolls-Royce Heritage Trust, Rolls-Royce plc, Moor Lane, PO Box 31, Derby
DE24 8BJ

Printed in 2003 by **océ** Printing for Professionals.

Océ Document Services, Bristol.

CONTENTS

		Page
Foreword		4
Preface		5
Chapter one	Early days	6
Chapter two	1943-53	16
Chapter three	Design of engines and accessories	29
Chapter four	T3 and T4, and twin-shaft engines	42
Chapter five	Rover Gas Turbines Ltd	54
Chapter six	Aeroplanes and auxiliary power units	62
Chapter seven	Boats and hovercraft	74
Chapter eight	Rover-BRM 1963-65	86
Chapter nine	Odds and sods	106
Chapter ten	Leyland and trucks	113
Chapter eleven	Is this the end?	128
Appendix A	Contracts issued to Rover 1940-43	134
Appendix B	Summary of the 1945 Rolls-Royce report	138
Appendix C	1950 RAC report	140
Appendix D	Comparison of compressor characteristics	142
Appendix E	Summary of major engine types	143
Appendix F	The Chrysler turbine car Sandy Skinner	144
Appendix G	The Fiat Turbina Andrew Nahum	151

FOREWORD

Following the famous swap of the Rover jet engine work (as recorded in our earlier volume *Vikings at Waterloo*) for Rolls-Royce's Meteor tank engine programme in 1943, Rover's interest in aero gas turbines effectively ended. However, after the end of the Second World War Rover built on this experience with a number of projects to develop small, cost-effective and efficient gas turbines, primarily – though certainly not exclusively – for road use. This book in our Historical Series records the reminiscences of some of the team involved in those projects.

Rolls-Royce at Derby had kept a watching brief on developments in the automotive gas turbine field from 1945 onwards: a cursory glance through the archives reveals a number of in-house reports and designs, as well as over a dozen patents specifically pertaining to road applications. The Company however, for whatever reason, decided not to enter what at the time other companies felt was a potentially large market. Interest finally subsided when Rolls-Royce declined a request from Leyland Gas Turbines Ltd to provide technical assistance in the development of small gas turbines for both road and rail application, in 1970, and events in 1971 focused attention on more pressing matters.

I am indebted to Mark Barnard for providing the manuscript of the Rover team's account of events, and his not inconsiderable patience whilst Ian Neish and I set about the editing. Illustrations are from the author's personal collection, unless otherwise credited, with a number from the British Motor Industry Heritage Trust (BMIHT) at Gaydon who also gave us permission to quote from H B Light's unpublished Rover history. Thanks also go to Trust members Sandy Skinner and Andrew Nahum for their unbiased appraisals of the Chrysler and Fiat turbine cars.

Richard Haigh
May 2003

PREFACE

This is a story. It is a true story about engineers within the British car industry 'doing their bit' from the 1940s onwards. Thanks to the imaginative management of the Rover Company in that period, a great deal of pioneering engineering on small gas turbines was embarked upon and the work created a lot of 'firsts'. Other companies sometimes did the job better – but at a later date and usually at much greater expense.

The idea to write the book came after reading the Rolls-Royce Heritage Trust book (No 22) entitled *Vikings at Waterloo* by David Brooks. The description of the way Rover became involved in Sir Frank Whittle's initial work in aero gas turbines, up to 1943, stimulated me so strongly that I asked David if he would also write about all the subsequent Rover turbine work. His answer, friendly but determined, was "No!". He then said that clearly the best person to write it was me. From 1953, I was in that Rover group of people and knew many of them personally, was interested in getting it all in print, and so had better get on with it. After the shock had passed, I had to admit he had a point and I have, in fact, enormously enjoyed doing it.

The total number employed on small gas turbine work by the Rover Company and its subsidiary, Rover Gas Turbines, amounted to 250 or so over a period of 30 years and it is from some of these, whose names are listed on the frontispiece, that I gleaned much of the information here published. Whilst turbine history, design and events feature strongly, a number of amusing stories have also been included as they reflect well the general atmosphere that existed amongst those involved. I have worked in the main from tape-recorded conversations with these colleagues and, having first not known what to say, they needed only a question about when they originally came to Rover to prompt a flood of memories lasting for the full tape or, in one case, for six whole tapes. Also, several of them have written (or re-written) parts of a chapter when their memory has surpassed my own.

This book is intended as a tribute to a group of engineers whose praise was rarely sung – more usual were reprimands for getting home late, spoiling the weekend or even, occasionally, being out all night.

My earnest thanks to those who contributed, and to my wife for typing and editing.

PISTONS TO BLADES

CHAPTER ONE

Early days

"Mark, you're searching a bit for the initiation of Rover's interest in gas turbines". These were the words of Bernard Smith (AB to us), Director of the post-war Rover Company, when I interviewed him for his memories of those heady days of 1940-43 when the Company first became involved with gas turbines. Rover's involvement with this novel engine started in January 1940 when Maurice Wilks, then Chief Engineer of the Rover Motor Company, met Frank Whittle for the first time. Having become frustrated with the progress of the work he was guiding to create his first jet engine at British Thomson Houston (BTH), Rugby, Whittle had been put in touch with Wilks through a mutual friend. Wilks was interested and it was suggested that Whittle's small company, Power Jets, might arrange for the Air Ministry to place some sub-contract work with Rover. In fact, direct contracts were awarded to the company. Things happened quite quickly then and a team of about six drawing office staff was set up at the Rover factory in Helen Street, Coventry.
The 'phoney war' was coming to an end as Germany started bombing British cities. Rover, deciding that things were getting too hot in Coventry, was moving the engineering staff to a requisitioned site at Chesford Grange Hotel, midway between Leamington Spa and Kenilworth. This was largely completed by a Saturday in November 1940, with work planned to start on the following Monday. Amazing timing as on the Sunday, in Coventry's Blitz, Rover Helen Street was destroyed.
  Jack Swain was the engine man for Rover and in 1938 had designed the sloping head engine, which was to see the light of day in the post-war Rover 75. As he went to work that Monday morning in November 1940, he saw that the Blitz had destroyed the works in Helen Street, so went on to Chesford Grange, giving lifts to people trooping out there on foot. He was installed in the ballroom, and turbine design work started. He had already spent two or three weeks at Power Jets in Lutterworth to be briefed on what it was all about, and had seen an engine test bed with its pitiful amount of equipment. Rover prepared a small shop in which to assemble engines and also planned a test facility there, using equipment salvaged from Helen Street. Ces Bedford recalls that, when he joined turbine design in 1941, the Drawing Office occupied the ballroom, while the drawings print machine and the canteen were in a cellar, alongside suits of armour.
  Rover's production director, Geoffrey Savage, had wisely instructed his buyer, Bernard Smith, to find some premises in the northern half of England to carry out this turbine work well away from the dangerous Coventry area. Having been told

not to come back to the Midlands until he had found an available factory, Bernard located an old cotton mill, Bankfield Shed, in Barnoldswick, Lancashire, which had enough space to become a manufacturing plant for the gas turbine parts. By the end of 1940 a further building, Waterloo Mill, was located in Clitheroe, a few miles west of Barnoldswick. Both were requisitioned by the Ministry of Aircraft Production (MAP) for Rover to develop, manufacture and assemble the components of the turbine engine and proceed with testing them and the complete engines. At this time, Ces Bedford was working for Ron Righton, the Rover works architect, and they went up to Barnoldswick to plan the conversions of the mills for this new turbine work.

Having past experience of their skills, Rover sub-contracted the fuel and combustion system to Lucas and called in a number of sheet-metal workers from Birmingham to help manufacture the ducting and combustion chambers. A small number of Rover and Lucas people moved into Clitheroe while the factory was being converted. By Easter 1941 more staff from both companies had moved in and were complemented by a growing local workforce. It soon became apparent that Lucas needed a lot more space for their sheet-metal work than had been anticipated. By the end of 1941, the Lucas people and their manufacturing had been moved from Waterloo Mill to yet another disused textile mill, Wood Top Mill in Burnley, which had also been requisitioned, for their specialised work on combustion chambers, sheet-metal ducting and fuel spray nozzles. Rover also chose Lucas for their experience with diesel fuel pumps and controls gained in Lucas-CAV at Acton, London, to develop fuel controls for start-up, acceleration and speed control, as well as altitude compensation for this novel engine.

The Drawing Office staff was increased from the original six men to a total staff of fifty and eventually, by 1942, the engineering staff totalled two hundred and fifty. Everybody worked six days a week plus overtime and, with enthusiasm and no trade restraints, things happened quickly. As an example, the aerodynamically straight-through B26 design was produced in ten weeks from receipt of drawings. Rover was given a number of development contracts from the MAP between 1940 and 1943, the first of which was received on 4 August 1940. They are listed in Appendix A and it will be noted that, for security reasons, complete turbine engines are coded as 'superchargers', ie centrifugal compressors for piston engines.

The abortive attempt to build a test area at Helen Street, and Rover's thoughts on creating one at Chesford Grange, were bypassed by the MAP. They decided that Rover Engineering Department should carry out the design work for a floating raft structure to measure engine thrust directly. This would be built at Waterloo Mill. The first two test beds were commissioned and operated by Jack Swain at Waterloo Mill in 1941 and continued to operate throughout the war. In the twenty months while Rover staff were at Waterloo Mill and Bankfield, a total of 1,200 hours of engine testing was recorded.

Ron Hill, a Rover design apprentice who became a member of the turbine design team and later Chief Designer, Gas Turbine Department, remembers that during 1940 not only did Frank Whittle occasionally come to the design office in RAF uniform, but Maurice Wilks and Robert Boyle (Assistant Chief Engineer, Rover) often called on Chesford Grange, sometimes riding their bikes to get there, to observe the design progress of the turbine unit by walking round the drawing boards and discussing the detail with the designers and draughtsmen. Everyone was very interested and discussions were lively and stimulating. The relationship between Wilks and Whittle was still amicable at that time, though one of Whittle's chief concerns (in addition to that of the award of direct contracts) was the copying of all Power Jets drawings by Rover. Rover quite rightly argued that, if they were supposed to get all the parts manufactured in secrecy, having Power Jets' name on the drawings was not the cleverest thing to do.

Another young man in the team was Adrian Lombard, who up to that time had been designing car suspensions with Gordon Bashford and had taught maths at the local technical college in 1939. He was cajoled into putting on a maths course for Ron Hill and others, as they were missing their expected training due to this move for turbine work. Unfortunately, they couldn't get enough 'customers' so the course had to be closed down, probably because there was insufficient finance – even in 1940. Norman Cooper, the gas turbine buyer from this stage onwards, had moved to Chesford Grange in 1940 with Gordon Bashford, Tom Barton and Norman Bryden, Rover section leaders who were involved in those early days of design. Designer Joe Brown joined at Clitheroe and Charlie Hudson, 'our' foreman, was also part of the team. Jack Swain remembers an occasion when a rather cocky Ministry official insisted on standing by an engine on its test bed, disregarding very clear instructions that no-one should do that until the engine had been run up to 16,500 rpm, to ensure its components were reasonably safe. This wise rule had been imposed after an impeller burst and had nearly killed two testers who were watching for fuel leaks. Jack decided to emphasise the cautionary note and ran the engine up to half speed and then caused the compressor to surge – not too difficult in those days. This would have been a terrifying experience to the uninitiated and, thereafter, the gentleman gave no further trouble! On another occasion, Mr Richard Ifield of Lucas was in a test cell trying unsuccessfully to get a fuel accumulator to release its fuel and Jack remembers Ifield holding his hands up in desperation. As he turned his back on the accumulator it chose that moment to release all its fuel, spraying him all over along with everything else. No real damage was done – but the story is typical of fuel systems development work.

Whilst everyone was working hard to produce the Whittle-designed reverse flow engine, Rover became keen on an alternative design of the engine internals, which would significantly reduce the very hot (excessively for the materials then available) sheet-metal duct work feeding the combustion gases to the turbine. This was the ST, or 'straight through', arrangement of the gas flows but it necessitated lengthening the turbine shaft and the distance between the compressor and turbine bearings. Whittle had considered this but had rejected it as it might incur serious shaft whirling problems and consequent engine failure. Perhaps, more importantly, the straight flow design would require a different combustion arrangement, a critical aspect of the overall design (and one which Frank Whittle was less confident of). However, Rover decided in mid 1941 to explore this further and Adrian Lombard designed a subtle way of splitting the main engine shaft into two, and rejoining the halves with a coupling sleeve over a spherical joint and all mounted in a third bearing. Apparently rather complex and expensive to make, it avoided the whirling problem, so opening the way to bypass the high-loss reverse flow arrangement of Whittle's design. This is known to this day as the Lombard Coupling. Another first was the Rover development of the electric starter motor (just like that fitted to a car).

MAP gave Rover an additional contract to design, build and test this engine, a task which was completed in nine months. The two engines shared the same aerodynamic components and were developed in parallel, the designations being W2B for Whittle's unit and W2B/26 for the Rover straight-through (ST) design. It is interesting that, in due course, the two designs went on to be developed into Rolls-Royce engine types; the Whittle design became the Welland and the Rover W2B/26 became the Derwent I.

Rover design W2B/26 – straight through combustion: general view (RRHT)

Rover design W2B/26 – straight through combustion: rear view – exhaust unit removed showing accessibility of turbine   *(RRHT)*

An early, and rather hazardous, remedy to engine problems was to fit a bellyband round the casing joint lines to reduce or stop air leakage. Jack Swain recalls one occasion when he didn't have the heart to ask anyone else to do it, so went into the test bed himself whilst the unit was on endurance, climbed up a step ladder and fitted this bellyband round the B26 unit casings. Brave man!

Another issue concerned carbon fouling of the ends of the fuel nozzles. Jack noticed some small spiral carbon marks, which suggested a strong swirl of the spray nozzles cleaning air. He deduced that the swirl was so strong that soot got sucked back over the end of the nozzle instead of being blown forward. He tried some simple vanes to prevent this and, thereafter, nozzles stayed clean.

I shall quote H B Light quite frequently, from his unpublished Rover history – *The Rover Company Limited and the Gas Turbine for Automobile Propulsion:*

> "During the twenty months when the Rover Company's Engineering Department was situated at Waterloo Mill, they:
> * Built up an Engineering Staff from about 20 to over 250 strong.
> * Carried out over 1,200 hours of engine test running.
> * Drew up in detail, and developed, a jet engine up to a thrust for take-off of 1,450 lbs and a test bed endurance of 100 hours.

* *Designed and manufactured a completely new layout of engine and proved it to have a performance of 1,450 lbs thrust, backed by three 50- hour periods of endurance running.*
* *Designed and manufactured all the engine Test Houses and various other rigs required for this work.*
* *All the work was carried out on a type of propulsion unit about which there was little or no previous knowledge and everything, (both in connection with the unit itself and test equipment), had to be developed from scratch; the final design which was evolved by the Rover Company forming the basis of a design for engines to be manufactured for large scale Service use."*

The first flight of a British gas turbine engine occurred on 15 May 1941 at RAF Cranwell. The test vehicle was a Gloster's purpose-designed and built research plane, the E28/39, with a W1 Whittle engine built by BTH (British Thomas-Houston of Rugby).

The following is the account of an RAF serviceman, Sid Hill, who later became a senior test engineer in the Gas Turbine department of Rover in the 1950s:

1939/40

"My involvement in the early days of Jet Flight was with the final assembly of the aircraft E28/39 at the Gloster Aircraft Company Ltd works at Hucclecote, Gloucester. The aircraft designer was Mr George Carter.

The later stages of construction were simplified with the use of a mock-up engine supplied by Power Jets.

Very close co-operation was maintained between Glosters and Power Jets in all of these operations; engine mounting, controls, instrument runs, jet pipe suspension, fuel system runs, and numerous other items were completed before the mock-up engine was returned to Power Jets.

At Glosters, the finals of the airframe assembly proceeded under much more pressure. Hydraulic and pneumatic systems have to withstand very high pressures for long periods, instruments and electrics function tests completed, undercarriage and brake function test schedule run off, flying controls, rigging geometry, control surface movement tested. All of these items must be documented and approved.

The W1X engine was received from Power Jets and installed. Engine runs were cleared by Power Jets' engineers and throttle stops set to 13,000 rpm.

April 1941

This engine speed will only move the aircraft at 20 mph on a grass field. The engine speed was increased to 15,000 rpm and this increased the aircraft speed to 60 mph. Taxi trials proceeded within this speed range.

A further adjustment within the fuel pump gave a final engine speed of 16,000 rpm.

The Pilot, Flt Lt P E G (Gerry) Sayer, made three fast runs and was airborne for 200/300 yards. There was, however, some reluctance for the tail to go down without some elevator assistance, the tail-skid touched the ground on one of the runs. I well remember Air Vice Marshal Linnell slapping my back saying, "Laddie, we have just witnessed the greatest step forward since the Wright brothers first flew". He was quite excited.

When the taxi trials were completed, the aircraft was taxied into the hangar and the detailed inspections began. The W1X engine was returned to Power Jets. The undercarriage was returned to the manufacturers; wheels and brakes to the manufacturers for inspection and clearance for flight. The airframe structure was inspected and cleared for flight – there were no problems.

The rear half of 'Regent Street Garage', Cheltenham, was taken over and sealed off. The airframe, wings and all components were collected here and the Aircraft E28/39, W4041 was re-assembled and prepared for pre-flight tests.

A full pressure test schedule was carried out and approved under-carriage tests were extensive and observed, as were all the pre-flight tests. The whole of this work was under maximum pressure and went on day and night.

On completion of the work, two articulated aircraft-transporter vehicles were left in the yard outside. The W1 engine was received from Power Jets and installed for flight. The flight crew backed the vehicles into the workshop and loaded the aircraft, with all the ground servicing equipment, during the night. A plywood airscrew was made and tied to the front of the fuselage, then the whole machine was sheeted over – it had the look of a fighter in transit.

From this point all security and transport was controlled by Scotland Yard. The vehicles were driven out of the workshop and the drivers never saw their load. We were all escorted in convoy; destination unknown until arrival.

14 May 1941

We arrived at Cranwell RAF Airfield. The crew promptly unloaded and assembled the aircraft for flight. However, weather conditions were unfit and P E G Sayer did some taxi work, testing a longer nosewheel to assist in take-off.

15 May 1941

It was evening before the weather improved and, at about 7.30pm, the first flight was away. It lasted for 17 minutes – the E28/39 reached 370 mph at 25,000 ft.

Fuel used	-	*Pool Paraffin*
Lub oil	-	*DEF 2001*
Test Pilot	-	*Flt Lt P E G Sayer (Chief Test Pilot, Gloster Aircraft Company)*

The first 10-hr Test Flight Programme was run off in 14 days without incident. No lubricating oil was added during the flights. The engine covers were not removed once during the flight test programme.

21 May 1941

This was a day in the programme which was allotted for a demonstration flight for a large number of VIP visitors: – Secretary of State for Air and Under Secretary of State for Air, Sir Geoffrey de Havilland and his son Geoffrey; Major Halford (engine designer); Mr C. Walker; Mr Roxbee Cox (MAP); Air Vice-Marshall Linnell; T O M Sopwith of The Hawker Siddeley Group and numerous others, including Group Captain Frank Whittle.

P E G Sayer took off at 6.15pm and gave a most impressive display with high speed, low level runs and vertical rolling climbs. The weather held off until Sayer landed; then it began to pour with rain. The event was very successful.

After Cranwell, E28/39 was moved to Shenington (Edge Hill) for further testing with W2B engines in an intermittent manner.

October 1941

The WIX engine, used for E28/39 Taxi trials was flown to USA in a Liberator to General Electric Company, Lynn, Mass, with a team of three Power Jets engineers; a complete set of drawings of W2B was sent by sea.

circa January 1942

At this point, test programme with E28/39 was slowing due to slow engine deliveries of W2B for flight. E28/39 was handed over to RAE Farnborough.

Much more emphasis was put on preparation for development test flights with F9/40 aircraft, which later became the Meteor. The flight crew was setting up the aircraft in a structure of girders with hydraulic jacks and micro-clocks to measure deflection under load of wings, fuselage, tail unit, etc. Finally the fully rigged machine was installed in a 'shaker rig' for extended vibration tests (day and night) at RAE Farnborough.

RAE was now running a test programme with (the second) E28/39 with a W2B engine. The pilot was Squadron Leader Davey. At 30,000 ft the temperature was -40° and the aileron-spar bearings became seized solid, making the aircraft uncontrollable. The aircraft carried out some strange manoeuvres as its height reduced and as it approached ground level, assumed level flight in a shallow dive and crashed on Laffins Plain, not far from our hangar. We rushed to the scene with equipment but found an empty cockpit. After some delay, Squadron Leader Davey came down on his parachute. He was rather distressed, having lost his oxygen mask and one glove when he bailed out. He put the pipe from the parachute oxygen bottle into his mouth and adjusted the supply with the shut off valve – at temperatures of -40°, not a pleasant experience!

December 1942

The Aircraft F9/40, DG206, was collected from Gloster Aircraft Company at Bentham, Glos. One of the prototype Meteors, in fact the first Meteor to fly, this aircraft had a special 'centre section' to take the larger de Havilland H1 engines. These were installed but the first flight was delayed for a modification."

The first unit from Rover at Chesford Grange was despatched to Power Jets in October 1941 with another two around April/May 1942. Considerable compressor surge was experienced, probably aggravated by the pressure losses in the reverse-flow ducting but, by de-rating to 1,000 lb thrust, two were delivered to Glosters in May and June 1942 for the taxiing trials. Runs in July 1942 confirmed that power was insufficient for safe flight but some flight trials were considered essential for this new engine so a Wellington bomber was fitted with a W2B in the tail gun-turret position. The installation completed 10 hours flying at up to 20,000 ft by September 1942.

The Rover-designed engine was running well in autumn 1942, though with excessive oil leakage. However, this did not prevent the unit being run to its full speed of 16,500 rpm and completing three 50-hour endurance runs by January/February 1943. Rover-built W2B engines cleared 1400 lb thrust for 25 hours and, in January 1943, a unit of 1,250 lb thrust was installed in the Gloster E28/39 and flew successfully on 19 March. Two more W2B units were up-rated to 1,600 lb thrust in May 1943 and fitted to the F9/40. Later, Rolls-Royce commended the work done by Rover, as the B26 engine was much less prone to compressor surge and power developed more quickly during the test programme. Rover management had an expression, which described their general attitude well: an open umbrella, resting on the ground with its handle pointing up, "If it doesn't work, it's no good!"

By 1942 Ernest Hives of Rolls-Royce was convinced that gas turbines were the future for the aircraft industry and later in the year Rolls-Royce took over all aero gas turbine work, utilising their 'air-mindedness', which Rover did not have and Rover gave up all their aero-turbine work and their parts of the Clitheroe and Barnoldswick factories to Rolls-Royce. In return, Rolls-Royce gave Rover their Nottingham factory to build the Meteor tank engine, based on the Merlin, enabling us to "grub about on the ground" – as Hives put it (with warm concurrence from Maurice Wilks). And that was the end of Rover's involvement with the aero-engine world – apart from some exploits, which are recounted in Chapter Six.

Ron Hill, like other Rover engineers, had been given the chance to stay 'up North', working on aero-engines for Rolls-Royce, or return to Coventry and resume his initial training in car design. He chose to return though, for a while, he helped on the conversion of the Meteor to diesel fuel. He then went to Chesford Grange, working on Rover's thoughts for new, post-war car designs. This was 1943 and Maurice Wilks came over quite often to direct and encourage. Ron liked and respected Maurice Wilks and calls him "a gentleman engineer". Another man employed there for a while was Peter Berthon, whom we will meet again later in this story.

Following the exchange of factories, Adrian Lombard elected to stay with the jet engine, and is pictured here (left) with Frank Whittle and Chief Test Pilot, Geoffrey de Havilland, after the first flight of the Vampire in 1943.

CHAPTER TWO

1943-1953

When Rover's involvement with aero gas turbines ended in 1943, Maurice Wilks and his brother, Spencer Wilks, Managing Director of Rover, were planning the re-establishment of Rover cars after the destruction of their pre-war Coventry base at Helen Street. The Government had built a shadow factory on Lode Lane, between Solihull and Sheldon, where Rover people had been building, repairing and testing RAF aero-engines; Armstrong Siddeley Cheetah (600 hp), Bristol Pegasus (1,200 hp) and Hercules (1,800 hp) – all radial piston designs. At the end of the war this factory became the main Rover plant. There were subsidiary factories at Tyseley, Acocks Green, Ryeland Road and Springfield, which provided the many technical services needed for engineering work and which had all played their part in wartime, but Lode Lane became the headquarters of the Company.

However, due to the Wilks brothers' considerable interest in the gas turbine engine, they had considered the application of this new prime mover in fields outside aircraft and, particularly, in road vehicles. It was not long before they pioneered the world's first design of a smaller engine of more modest power. They also adopted two major features, necessary for road use. The first was the twin-shaft arrangement, where the majority of the thrust of the jet exhaust is harnessed to drive a second turbine, called the power turbine, which is mechanically independent of the compressor and turbine of the simple engine. This power turbine is geared to drive the output needed, wheels in a car, or propeller in a boat, for example.

On leaving the power turbine, the engine exhaust has lost most of its pressure but little of its heat, so the second feature was the need to design a heat exchanger which would remove as much as possible of the thermal energy in the exhaust and re-insert it into the engine airflow just before the combustion chamber. The heat-exchanging would prove very difficult to achieve effectively at a sensible cost and for an acceptable life and you will see how it developed through our story.

In 1945, a nephew of Maurice Wilks joined Rover, having served an apprenticeship on gas turbines with Rolls-Royce at Derby. This was Charles Spencer King, known to everyone as Spen King. He brought with him a colleague from Derby, Frank Bell, a New Zealander who became known as 'Tinkerbell'. Another early member of the turbine team was Harry Knowles, a bachelor and an unusual personality. He was meticulous with his calculations, stressing or form development, always throwing away the figures when the task was complete – he said, "there was no comeback that way!" The team was originally sited in a corner of the Rover (car) design office and every morning half-an-hour's discussion took place between Frank and Harry on any topical subject, not necessarily gas turbines. Several times, it is reported,

Harry arrived in the morning saying he had fallen in love with somebody he had seen at the bus stop, but he never did anything about it.

At this time Rolls-Royce was also considering automotive gas turbines. Report reference Lov/FRB2/JF, *The possibilities of the gas turbine, a motive power for the automobile,* dated 2/3/45, the summary of which is copied as Appendix B and outlines the advantages and disadvantages for a 90 hp engine, based on a twin-shaft design with heat exchangers. A 5" diameter centrifugal compressor running at up to 70,000 rpm was scaled from Merlin experience in performance, bearings, lubrication and gearing and none of these aspects was considered to give cause for alarm. More surprisingly, no difficulty was expected from the heat exchanger design, manufacture or life. The report has much to commend it and it is interesting – despite its conclusion – that Rolls-Royce management decided not to explore this field at this time.

In 1946, two important issues occurred. First came the move of the embryo turbine drawing office from being within Rover Design down to Project A – the rather boring long bungalow building remote from the Lode Lane end of the factory. Frank Bell went first, followed a week later by Ron Hill and, another week later, by Ces Bedford. Charlie Hudson started up Project B, a building just a few yards beyond Project A, which became an engine build area, with some large iron-doored cubby holes which took on the role of rig shops. The technical office was in the same building as the drawing office, in which Harry Knowles (stressman), John Garrett (turbine design), Phil Phillips (development) and Fred Pickles (development) operated. Ray Bourne, suffering from a limp, also joined in Project A as there were no stairs to go up. We started a tracing group with Molly Barker and Thelma Hands – we were growing up. Ces Bedford remembers working till midnight (note – nothing to do with Molly and Thelma) once or twice and then walking home to Stoke, Coventry – a distance of over 10 miles.

The second event was the first product from the small secret team called the T5 unit, this referring to the diameter in inches of the centrifugal compressor of the engine, and targeted at 100 hp. Both turbine rotors – the first being part of the gas generator, which drove the centrifugal compressor, and the second, the power turbine – were of 'axial-flow' type, i.e. they were discs with blades on the outer rim. These turbines had integral disc and blades machined from one large solid chunk of a Nickel alloy, Nimonic 90. Air-blowing tests were carried out to check the performance of the turbine designs on a rig, which used a large amount of high-pressure air from a Rolls-Royce Merlin supercharger, driven by a 500 hp electric motor.

After these turbine tests, attempts were made to operate the complete engine and a maximum of 30 hp was obtained for a few minutes and the longest run ever was 30 minutes. The design was plagued by tip rubs, distortion, and even overspeeding due to running away from the fuel control device. However, all the apparent negatives were excellent experience and it was decided early in 1948 to design a new unit with an 8-inch diameter compressor – i.e. the T8. A number of these engines were built and initially produced around 100 hp. They were more successful than the T5, but Lofty Poole (fitter/tester) recalls the occasion when, as Spen and Tinkerbell were watching, the unit on the test bed ran away to a high speed – i.e. it went on accelerating out of control. Lofty switched off but this had no effect, and Spen decided that to get away was the most sensible thing to do. Tinkerbell decided otherwise. He got his slide-rule out and reckoned we were getting 460 hp before the engine finally blew up. All he could say was "I can't believe it", several times. After this incident Spen proposed (ordered?) that a disc be kept available to blank off air at the intake – which would stop any engine if the need arose.

Also that year, the Rover turbine team cut the top off a P4 Rover saloon car and fitted a T8 engine in the space behind the front seat, driving the rear wheels. The car eventually carried the registration JET1. Naturally, there were many detail changes carried out on the engine and its installation as time went by and experience was gained with development testing. The initial engine was designed with two combustion chambers but unreliable ignition of the second chamber, causing long flames through the engine and out of the exhaust, led to redesign with only one chamber. It was successful. Another problem, well known to engineers in later days, was explosive light-ups when the engine was started and already warm from a previous run. In one case, a large square-shaped exhaust pipe was 'blown-up', nearly round, the resulting BANG causing considerable momentary panic. Rarely was any damage done to the engine, but bangs or surges have always been part of a turbine engineer's life and one cannot say that running one was dull.

Another 'interesting' event occurs when the combustion flame goes out whilst fuel is still being injected. This results in an enormous cloud of unburnt vapour, which can be seen and smelt for some distance round. Flame blow-outs will always feature strongly in the memories of all those involved in this story – we were not always very popular with the factory or, later, the public.

More from H B Light:

> *"In January 1949, the Press announced that engineers of two firms, working in secret in London and Birmingham, had reached an advanced stage in the development of gas turbines for automobile propulsion - the Rover Company, who developed the early aircraft turbine design of Sir Frank Whittle, and Centrax Power Units Ltd, of Acton, Middlesex, run by members of Whittle's old Power Jets Team. Work on the Rover project steadily developed following the announcement of Rover's interest in the gas turbine. The Company had no illusions and at no time did it think it would be able to make a satisfactory car gas turbine engine within a short time."*

At this time the Rover turbine team numbered 14 people of which only two, Lofty

Poole and Eric Eshbourne (fitter/testers), were allowed to run the engines on the test beds. Foreman Charlie Hudson proved to be a fascinating character. He could sleep anywhere, anytime. He could be quite fiery but got things done. He once sacked Lofty, sent him away with his tools – and then re-instated him a few days later.

Charlie Hudson, in typical pose, pictured here on the Torquil launch in Wales, 1949

JET1 was being used as a mobile test bed and the T8 engine was giving around 100 bhp at this stage. A regular test route was past the large test hangars and round their backs to Project A and B. Occasionally, a flame blow-out occurred and enough smoke came forward into the cab to necessitate a small two-inch high perspex 'screen' being mounted behind the seats, in front of the engine and its exhaust, to avoid this blow-back embarrassing the driver and passenger (as well as half of Solihull).

Quoting again from H B Light:

"It was thus that Wednesday, 8th March 1950, marked a new epoch in Automobile Engineering when the Rover Company submitted a car fitted with a gas turbine power plant for test at the Motor Industry Research Association (MIRA) proving ground at Lindley, in the Midlands.

The test was held under the observation of the Engineering and Technical Department of the Royal Automobile Club. This car, which was then unregistered, later became known by its appropriate registration – "JET 1" – the world's first Gas Turbine Car.

On the following day, 9th March 1950, "JET 1" was transported to the Silverstone Race Track for a demonstration for the Press, BBC Cinematographers and Officials of the Society of Motor Manufacturers and Traders and many well-known personages of the scientific and industrial world.

Rover's achievement received worldwide acclamation and publicity under such headings as:

'British Motor Car Company unveils World's First Jet-Engined Automobile'

'Gas-Turbined Car has World Premiere – Rover Company Makes History'

'Britain's new Jet Car astonishes Foreign Experts'

'Made in Solihull: World's First Gas-Turbine Car' etc.

And, of course, in some foreign papers there was a mixture of opinion and some scoffing at the British achievement. A report from Detroit suggested that the Jet-Propelled Automobiles would make toast of pedestrians and burn other cars right off the highways, and they talked about people being 'seared to a crisp' if they stood behind one when it started up. This type of thinking was no doubt due to the car having been erroneously called a JET propelled vehicle, whereas in fact it is propelled by a gas turbine engine – quite a different matter!

Privileged onlookers at the two demonstrations were certainly thrilled at seeing the shapely and well finished two/three-seater car showing speed and ease of handling. With two pedal control - accelerator and brake pedal only – and no change speed gearbox or clutch to consider, the car moved smoothly forward as the accelerator was depressed to provide enough energy in the gas stream to rotate the power turbine after the manner of a fluid drive. Starting was prompt, since no warming up process is necessary as there were no sliding frictional services to be lubricated."

The turbine car, later known as JET1, at the RAC Test in 1950, with Maurice Wilks, Bernard Wilks and Frank Bell. The RAC test report comprises Appendix C.

JET1 was also the world's first car to use disc brakes – we worked with Girling and Dunlop on these. It certainly needed them because of the lack of over-run braking from the engine.

On one test, Lofty remembers he was head down in the cockpit reading temperatures of the brakes and their fluid and Spen, who was driving, didn't believe what Lofty was reading and he also put his head down below the windscreen. Not surprisingly, there was an almighty BANG as the car struck a sand-filled barrel acting as a marker and, unfortunately, the brake calliper was bent. In a later run with Peter Wilks (another nephew of Maurice Wilks) driving they used the runway, which goes across the circuit and perimeter track to get some speed up. Peter came tearing up to the group of engineers waiting and, as he passed, put both his hands up in the air – as if to say "What can I do now?" – he had boiled his brake fluid and, because of the bent calliper, the car would not stop. He flew across out of the perimeter track, getting through a gap between two markers with an inch to spare, and found enough space to let the car slow down of its own accord on the apron. Another close shave! Fluid boiling was a regular problem before Girling developed a special fluid with high boiling point.

Type T8 200 bhp twin-shaft automotive engine with twin combustion chambers

A further quote from H B Light concerns 1951 and 1952:

"The Dewar Trophy Award – The Rover Company's long record in automobile engineering had earned it universal respect and its enterprise in gaining for Britain the honour of being the first country to introduce a gas turbine car is not easily measured. It was, however with widespread satisfaction in motoring circles that the Royal Automobile Club decided to award to the Rover Company the famous Dewar Challenge Trophy for the most outstanding technical achievement of the year.

In some respects, the Rover turbo-car could be considered the greatest achievement of all, for it represented a challenge to the strongly entrenched piston engine.

The Dewar Trophy award to the Rover Company in 1951 was the first award of the Trophy for 22 years. (It was last awarded in 1929 to Miss Violet Cordery for a 30,000-mile run on Invicta chassis at Brooklands.) This award to the Company was the second in its history – the previous one being in 1925, when a 14/45 hp saloon model made 50 consecutive ascents of Bwlch-y-Groes Pass, North Wales.

Presentation of the Trophy was made at a lunch given by the Royal Automobile Club in London on 20 April 1951.

Amongst those who co-operated with the Rover Company in the design of the turbine should be mentioned Henry Wiggin & Co Ltd, manufacturers of the Nimonic alloy (a Mond Nickel patent) high-temperature-resistant turbine wheel and blades, machined from the solid; Joseph Lucas Ltd, who were responsible for the combustion system and electric starting, and Shell-Mex and BP Ltd, with Ricardo and Co Ltd, who assisted in research.

... Early in June 1952, the Company announced its intention to send the car to the Continent to undergo technical and speed trials in an attempt to set up a standard world speed record for jet cars. It was planned to exceed 100mph and the track chosen for this important test was to be the Jabbeke motor road near Ostend in Belgium and would be observed by the Royal Automobile Club and other International motoring officials. The Rover Company was anxious to show that Britain still held the lead in this particular field of motoring and success would mean considerable enhancement of British Engineering prestige.

The sleek grey-blue Rover JET 1 began warming up early on the 25 June 1952, ready for the traditional speed tests of the flying start and standing-start mile and kilometre.

Since, for these trials, fuel consumption was of secondary importance, no heat-exchanger was fitted. There were many detailed mechanical and metallurgical developments incorporated in the new engine, resulting in the output of the engine being raised to 200 bhp, considerably more than the output of the original engine.

The honour of recording the first land-speed record for a gas-turbine driven car was to be shared by two drivers: Mr Spencer King and his cousin, Mr Peter Wilks, nephews of Mr S B Wilks, Managing Director of the Company.

During the first day's trials, another milestone was passed in the development of the world's first gas turbine car. The Rover JET 1 which, at Silverstone track two years earlier achieved 85 mph, reached a speed of more than 140 mph on the Jabbeke motorway between Ostend and Ghent. By the second day the official speeds logged with the International Automobile Federation were:

| Flying kilometre | - | 151.965 mph |
| Flying mile | - | 151.196 mph |

Full details of the car times were :

Flying kilometre	-	14.65 sec (152.691 mph)
and	-	14.80 sec (151.143 mph)
average	-	14.72 sec (151.965 mph)

Flying mile	-	23.92 sec (150.000 mph)
and	-	23.71 sec (151.833 mph)
average	-	23.81 sec (151.196 mph)
Standing kilometre	-	26.97 sec (82.941 mph)
and	-	27.33 sec (81.848 mph)
average	-	27.15 sec (82.391 mph)
Standing mile	-	38.20 sec (94.240 mph)
and	-	37.07 sec (97.113 mph)
average	-	37.63 sec (95.668 mph)"

The 1952 Rover Turbine team just prior to the author joining the company. Spen King is fourth from left, and Frank Bell fifth from the left on the front row. *(BMIHT)*

Spen King at the wheel of JET1 in 1952. *(Motor)*

The trip to Belgium had its lighter moments too. The Customs man there was most disturbed that he could find no engine number on the engine though he had been told the number verbally. Arguments that this was the only gas turbine car in the world availed nothing, so Lofty got out his toolbox and scribed the number on an appropriate casting of the engine – and the Customs man was satisfied. The most serious technical problem encountered involved Lofty having to replace a gasket on the oil pump. Spen impressed the motoring writer Laurence Pomeroy when he gave him a lift in JET1 after some high-speed runs. Spen was in a bit of a hurry to cancel some insurance and when Laurence returned, he told Lofty he had seen the speedometer reading 154 mph!

A further tale for 1953: Finland was quite an important export country for Rover, and Mercedes Benz were pushing their noses in, so Rover decided to fly the flag by sending over JET1. Helsinki is the capital city and demonstrations had been arranged around a small race circuit nearby. Spen drove, of course, in a fairly enterprising way and Lofty noted that people were timing him round the circuit. At one stage, Spen went sideways but kept it 'on the island' and we heard that his fastest lap was the fastest of the day in the race meeting. That same year JET1 (plus, I believe, two spare power units) was retired off to the Science Museum in London to have a well earned rest.

There were three applications of the T8 engine which interest us – the car JET1, a fast launch Torquil, and 4 engines bought by the Government Naval Research Laboratory.

The second Rover gas turbine venture, in 1950, involved Torquil, a 60 ft long launch with two 250 hp Dorman diesels, which belonged to Spencer Wilks. He wanted to fit gas turbines to demonstrate to the engineers in our Government, particularly Mr Roxbee-Cox, Chief Scientist at the Ministry of Fuel and Power, that Rover should get a larger allowance of steel, which was still heavily rationed, to build motor cars.

Spen chose Tony Worster (development) to design the installation of two T8 engines but to retain one 100 hp Dorman as a precaution. The turbines drove one propeller each and a third propeller was driven by the Dorman. At this stage the T8s were developing around 160 hp and, being so smooth, were mounted solidly in the hull. The boat was moored at Bangor, with Charlie Hudson and Lofty Poole as turbine crew. All the fittings for the installation were made by Rover. Two large fuel tanks fed kerosene to the turbines and three sets of batteries provided 24 volts for the diesel and 12 volts for the turbines. Came the day for the first run; turbines started up for a short cruise, one reversed to spin the boat round and return. A few days later, Spen opened up and 15 knots was measured, with jet pipe temperature (jpt) below 600°C – so all was well, though the boat would not plane on the water. Lofty Poole tells an amusing (with hindsight) tale of Torquil's engines. Strict instructions from Spen not to exceed 700° jpt during start-up were regularly ignored or the engines could not reach idling speed and, to help the research, Sidney Heslop (who became Chief Metallurgist of Rover) used a little eye gauge through which he looked up the jetpipe. Unfortunately, we had a wet start which sprayed out burning fuel and Lofty remembers rolling Sidney over on the ground to put out flames on his clothes! Torquil was well-silenced, though the exhaust smell from the twin combustion chambers was rather unpleasant.

The decision was made to demonstrate Torquil on the Thames to the former Chancellor of the Exchequer, Sir Stafford Cripps, which involved taking the boat overland from Birkenhead to be moored by the Chelsea Embankment. The engines were run to check their state but a turbine tip rub was experienced. Everyone knuckled down and changed the power turbine unit but this also had a tip rub. Off again came that unit and Tony Worster drove back to Solihull to pick up yet another power turbine assembly, getting back down to the boat by 2.30am. By 5.00am, a third tip rub had happened. In desperation, Spen took the assembly to the Segrave Road Rover agent, put it on a lathe and ground an enormous amount off the tips. Back at the boat at 9.00am, he tried to start the engines but the batteries were flat due to being used all night. So Spen hooked up the Dorman's batteries and, with everyone holding wires, spanners, etc, got the turbines going by 10.00am - for an 11.00am visit by the VIPs. Spen then stopped the engines because it was required that the visitors witnessed them being started. So aboard they all came and – wonder of wonders – the engines started and all was well. We did a little demonstration run for the Press, and repeated this the following week at the Motor Show. After showing the Torquil publicly, the turbines were removed from the boat and it was sold.

The Torquil launch powered by two T8 200 bhp gas turbine engines

The Admiralty Experimental Laboratory (AEL) at West Drayton bought four T8 engines and ran them on test beds, also fitting one on a harbour launch for Portsmouth. They tended to make carbon rather easily and hard bits would break off in the chamber and cause blockage in the nozzle vanes – causing the engine to surge. Noel Penny (development) recalls going down to West Drayton with Charlie Hudson to clean the spill burner and remove carbon. He also believes Ricardo helped us to change the engine from two chambers to a single one – as used on JET1.

In the early 1950s, two other projects were taken on by the post-war Rover management, which illustrate the technical enthusiasm of the Wilks family. Nothing to do with cars, but interesting technically, these elicited some funds from the Government. The first was a rotary-valve swashplate torpedo engine, working on HTP (hydrogen peroxide), a mono-fuel which carries its own oxygen. The Navy gave us the contract as they were exploring this fuel's use in submarines. It is rather beastly stuff, as I had demonstrated to me when a rag soaked in HTP was thrown down on ground, which was already wet with paraffin. A little puff - and the rag burst into flames! This engine was built round a large taper-roller bearing and Jack

Pickles, project engineer in charge, decided that he could drive some auxiliaries off the bearings' cage – ingenious. The engine was quite successful, though I recall a telltale patch of new bricks in the wall of a test hangar where an engine had 'sort of exploded' once. The contract was terminated when the Navy decided HTP was too dangerous to have around their ships.

Type T8 engine with single combustion chamber

Jack Pickles was also project engineer for the second project – a neutron spectrometer. That 'mouthful' describes (as much as we knew) a device where we shot a stream of neutrons through a slot and regulated the rate at which they passed through by varying the speed of the rotor. This was a project for Atomic Energy Research Establishment, Harwell and, we supposed, just a one-off. However, as it turned out, various Governments came to us to repeat the order. The testing was carried out in a brick-built bunker, near one of our test beds. It was a clever design, made in plastic material and mounted in rolling element bearings. You can't say the Wilks were not enterprising!

CHAPTER THREE

Design of engines and accessories

Whilst it was always the dream of the Wilks brothers to explore the possibility of selling cars powered by a gas turbine to the public, there were many aspects of this novel engine that needed better understanding first. Besides those aspects peculiar to cars (power levels, fuel consumption, noise, availability, etc etc) experience could be gained in a general public environment by selling simple gas turbine engines to do a job which other engines could not do so effectively. The Wilks were well aware that this was a way that Rover could start, as they had pressure from military test establishments who wanted an engine for applications in military service where the excellent power/weight ratio of a turbine left regular piston engines well behind. Fast starting straight to full power and multi-fuel capability were other assets. Hence a Rover small gas turbine was attractive to our Government in Britain after the Second World War and our Navy's interest in the first T8 engine has already been mentioned. The Admiralty research establishment at West Drayton, London, was always in the forefront of pressure for results and granted Rover a number of useful contracts to encourage the work.

In August 1952 we ran our next engine, T6, which became the 2S/100 or Aurora, with a six-inch diameter compressor. In addition, by omitting the power turbine section and modifying the first turbine, we had another prime mover – called Neptune – which later became an engine produced in numbers, the 1S/60.

It would now be helpful to consider what is a gas turbine, how it works, and how the basically simple design needs to grow and develop to become the desired car engine.

The first 2S/100 twin-shaft gas turbine, developing 100 bhp

Single-shaft turbine design

In its simplest form, the engine design comprises a centrifugal compressor mounted on the same shaft as a turbine rotor, which provides the power to drive it, plus an excess which becomes the output of the unit, plus power to drive the oil and fuel pumps and any other device, such as cooling fans or a generator. The airflow is drawn in from outside by the compressor, usually through an air filter, to the combustion chamber, into which fuel is injected for burning to raise the mixture to a high temperature, then down a volute into the turbine to produce power and, finally, exhausted back to the atmosphere. There is much literature about this turbine cycle so we will not dwell upon it, save where some change is being described as part of our history.

With our turbine, two means of starting were used for production engines – electric motor and hand winding. In the first, just as for piston engines in cars, a battery is connected to the starter motor, which causes it to turn at high torque. The turbine unit is connected via gears such that the electric motor will rotate the engine at around 15-20% of operational speed. As the speed rises, fuel is sprayed into the combustion chamber, where sparks are generated at an igniter plug and the fuel/air mixture is set alight. As the flame is initiated, more heat is released into the turbine, more power is generated and the electric motor plus heat energy work together to accelerate the engine further. Probably at 40-50% of operational speed, the electric motor is cut off and its drive is released through some form of clutch. From this point the engine will accelerate further until the fuel system's own controls take over to stop the engine accelerating beyond its operational speed – call that 100% governing. The gears connected with the turbine shaft now rotate the output shaft and power may be taken out of the engine. As power is taken, the fuel system will automatically feed in greater amounts of fuel to maintain 100% speed. Because it is so easy to overload a turbine and cause the temperature to rise to levels which would be dangerous to the materials used - particularly for the turbine blades - a form of temperature limiter is fitted which prevents excessive fuel from passing into the engine. If the overload is maintained this will consequently bring down the speed until it eventually stops. If the excess load were removed when first applied, the engine would continue happily running at 100% until the fuel valve is turned off to stop the unit. This description defines the simplest form of prime mover gas turbine engine.

If we consider the 1S/60 as fairly typical of such an engine, let us look at some of the relevant numbers, which will prove useful for comparisons with other prime movers, as well as the developments Rover undertook. 100% operational speed was 46,000 rpm, an order of ten times the speed of an equivalent power petrol engine for a car, and the shaft carried the first stage of a high-speed gear train. The design of high-speed gearing we used had been brought by Harry Knowles from Rolls-Royce. It operated on a 40° helix, which is steep, and caused difficulties in manufacture, entailing the buying of a Swiss MAAG grinder and developing a highly-skilled engineer, Bill Gregory. Later, MAAG hired our Bill several times to help them explain their equipment to other customers – so skilled had he become. Once the gear reduction has brought speeds down, such components as oil pumps, fuel pumps, tachometer drives, etc. can follow normal practice, around 4,000 rpm in the 1S/60. The drive to the power output can be selected to any desired value, in our case usually between 2,000 rpm and 5,000 rpm. Again, though the turbine speeds are so high, they are steady and smooth, not having any pulsations as with a piston engine, and are therefore gentler on gears, bearings and mounting structures.

Note also that, for a 60 hp output, the turbine section is producing around 200 hp of which 140 hp is used to drive the compressor mounted on the same shaft, and only the 60 hp passes through the gear train. This 60 hp is available at operational speed (100%) but, due mainly to the very 'peaky' output from the centrifugal compressor, there is a serious reduction in power available at lower speeds. For example, only about 20 hp is available at 75% speed and near enough 0 hp at 50%. This characteristic is acceptable for static installations such as electrical generators, water pumps and fuel pumps, which normally operate at one speed, though sometimes we had to use a clutch to permit the engine to start satisfactorily.

Twin-shaft turbine design

The power characteristics described above would be totally unsuitable for a road vehicle, which obviously spends a lot of its life near zero speeds and needs lots of torque at these low speeds to accelerate. So it was no secret that everyone exploring gas turbines for road use was working on twin-shaft engines. As its name implies, a second turbine and shaft are used, where the second is not mechanically connected to the first at all. The only connection is the exhaust gas from the first passing through the second turbine before being discarded back to atmosphere. In this case the first turbine unit, as already described, just produces lots of hot gas. In passing through the second turbine, the hot gas produces torque, which can, via gearing, be directed to the drive wheels of a car. At zero output speed (standstill) torque is about 2¼ times maximum power torque, so the power turbine acts as quite an efficient torque converter – and gas turbine powered cars often had no intermediate forward gears, just one for reverse. Some American cars used two or three forward gears to give their heavier cars better acceleration, and the turbine-powered trucks used four or five speed boxes (where their piston-engined rivals used ten gears). This is an attractive attribute for the turbine power unit, but has a penalty in increased fuel consumption if the gas generator is frequently extended in vehicle acceleration – as it usually was. This still relatively simple twin-shaft turbine was that used for the first gas turbine cars – the T8 engine in JET1 was of this configuration. However, there were three major difficulties to be overcome in order that a desirable motorcar might be produced and sold to the public. These are briefly addressed next and are - fuel consumption, response characteristics and cost.

JET1 was certainly a fast car, attaining just over 150 mph in controlled and recorded runs on straight Belgian autoroutes in 1952. The first bad news was, however, that this car, though easily driveable on normal roads and in traffic, returned between 1 and 4 miles per gallon on paraffin/kerosene fuel. The cause of the problem was the enormous amount of heat being rejected in the exhaust. One of the most difficult jobs we had was to develop and improve a device called the heat exchanger. Its purpose is to remove as much heat from the exhaust as possible and 're-insert it' in the engine, after compression but before combustion. Heat from the exhaust, at temperatures from 400° to 800°C, is transferred to the relatively cool high-pressure air from the compressor (up to 50 psi). Sounds easy, but it's not.

Briefly – two types of heat exchanger are practical – the recuperator and the regenerator. The former separates the two airflows mentioned already into flat tubes. Inside and outside the tubes are fins, usually corrugated metal, which are brazed to the walls. The cool, compressed air will be passed through the inside fins and the hot exhaust through the external finning. Heat is transferred from gas to fins, through the tube wall and into the air through the other fins. This is called a secondary surface recuperator and ours reached around 70% efficiency, which is the ratio of heat actually transferred to that theoretically available. I remember 'inventing' an inspection rig, which passed very cold liquid through the tube and blew warm air over the outside. Any fins that were not successfully brazed did not ice up and were easily observed.

The first heat exchanger inspection rig used to measure thermal efficiency of the elements. *(BMIHT)*

The first heat exchanger Rover designed was a secondary surface cross-flow unit, which is not an efficient design but simple to manufacture. However, there were insufficient supports in the flat part of the tube so, when Jack Belfit (fitter/tester) did a pressure test on them, they became large and round and hence useless. The next stage was the secondary surface contra-flow design, which was better than the cross-flow, and made it into our first engine unit and was fitted to the Aurora engine. It took us some time to get a reasonable heat exchanger and JET1 missed out on this – thus remaining rather a thirsty car.

The 2S/100 (Aurora) engine with heat exchanger.(BMIHT)

Later there was a development called the primary surface recuperator, where no fins were needed and the interface was itself corrugated like fins to obtain a large surface area. These were more difficult to make and seal to prevent high-pressure air leaking into the exhaust, but reached 80% efficiency.

The regenerator design had a large number of small passages with a big surface area (called the matrix) mounted in a big disc. This was rotated across the gas streams so that, at any one moment, exhaust gas gave up most of its heat to one half of the matrix, while compressor-delivery air was blowing through the already heated other half of the matrix. This is a continuous process because the big disc rotates slowly (typically 20 rpm), while the gas streams are constrained to pass directly through it, front to rear, by rubbing seals over the surface of the matrix. Because of the temperature difference between front and rear faces, the metal matrix disc coned considerably and, though this made sealing difficult, these attained 85% efficiency, with perhaps 3% of the high pressure air leaking away and a wear problem on the seal. This design was favoured by the Americans, whilst Rover initially developed the secondary and primary surface recuperators.

However, by 1965 Rover had decided to change to the regenerator as, by this time, a ceramic honeycomb disc had become available, and offered probably the best chance of attaining both performance and life at a cost which was significantly lower than any metal heat exchanger. To show where the ceramic heat exchanger took the gas turbine engine, we can quote the Rover BRM 1965 performance of 13mpg, against Ferrari and Porsche giving 10-15 mpg, at the Le Mans motor race. The same Rover BRM gave a best figure of 31 mpg at 43 mph and about 18 mpg on average road driving. So the problem was not solved, but more realistically balanced with the piston engine.

Primary heat exchanger exhibiting typical damage of cracked leading edges. *(BMIHT)*

The second difficulty, though not as critical as fuel consumption, was the engine response characteristics. You will recall how compressor surge, a momentary reverse of airflow in the compressor, usually with a mighty bang, caused a lot of difficulties with early engines, both aero and small. In car engines, surge is particularly aggravated as an engine must respond to every whim of the driver's foot on the throttle and Rover had to develop the compressor continuously throughout our research. Once, when Phil Gardiner (Development) was the heat exchanger engineer, he leaned over the engine to work on the heat exchanger mounted on top. The accelerator was pressed, the engine surged and Phil, terrified was up a grass bank ten yards from the car faster than he would have believed was possible!

Appendix D shows how we progressed in development from a Rolls-Royce Merlin supercharger to the ultimate swept-back impeller of Peter Parker (Development). The final compressor was remarkably good at avoiding surge in the engine, as well as being about 4% more efficient than early units.

The ultimate swept-back impellor, which was installed in the 2S/350R engine (see Chapter Ten).
(Peter Parker)

Cars like JET1 had no moving nozzle vanes and it took a significant period to bring the engine from idle speed to full acceleration – about 5 seconds, during which the driver had to be patient and, clearly, on normal road conditions this could be dangerous. Also, when the throttle was lifted, it took a few seconds for the gas generator to reduce speed back to idle and during this some forward torque was still being produced.

Early work with JET1 produced a number of incidents where the front brakes were hard on (through the foot brake) while the back wheels were still producing forward torque for several seconds as the engine slowed down – not a pleasant sensation. My first experience of work at Rover – as a graduate – was to help repair JET1 after Charlie Hudson had bumped it for this reason. Chrysler, and the other American researchers, developed variable nozzles. These were short vanes on the power turbine nozzle, which could be rotated very quickly to direct the gas backwards on the power turbine blading. This produced reverse thrust, with an effect not dissimilar to normal overrun braking on a piston engine, so making driving more acceptable and safer. They have small efficiency and cost penalties but these can be accepted. There was a further purpose for these variable nozzle vanes, which became very significant later. Apart from the heat exchanger, another way to improve engine efficiency, and hence fuel consumption, is to change the angle of those power turbine vanes in the forward power direction to alter the engine temperature. As the vanes are closed they create back pressure and the engine runs hotter and fuel consumption is improved. This will be discussed further when the Rover truck engine is described.

The Third difficulty concerns the cost. The gas turbine is indeed a relatively simple engine and is light, but uses a quantity of expensive materials because of the high temperatures involved. These materials involve alloys of iron with nickel, cobalt, chromium, plus some rare earth materials, such as niobium. Due to their scarcity and costly mining, they will stay fundamentally expensive forever. When gas turbines started in the 1940s, there was almost total ignorance about the potential performance of these high temperature alloys and as turbines became significant for powering aeroplanes, a national effort exploring their metallurgy was initiated. Also, the machining effort of these materials was extremely difficult and, hence, expensive. The evolution of precision casting, usually in a vacuum or in an atmosphere of inert gas, was also developed but it took until about 1955 for these material and manufacturing developments to become generally available. One has to say that today this cost problem is still not resolved. The piston engine, although now having become more expensive through turbo-chargers, intercooling, etc is still considerably cheaper than an automobile gas turbine. The only potential solution envisaged by some is to use ceramics as the material for the high temperature components, and work continues around the world to this end. More about this in Chapter Nine.

One or two more complicated designs of gas turbines have been, and maybe still are, being explored, which depart from the relatively simple twin-shaft turbine configuration described. Power turbines have been mounted before the gas generator turbine, a third shaft has been added to split up the compressor and turbine stages, but none of these is attractive enough to be the 'breakthrough' for automotive gas turbines. I have always liked the expression KISS – 'Keep It Simple, Stupid!' – and maybe it's not so stupid for the gas turbine to stay that way.

T8 engine had used a spill-type burner, which had a lot of fuel pumped into it, quite a lot recirculated, and the difference was the sprayed fuel. This technique gave the fuel kinetic energy to swirl around in the atomiser, helping to produce a fine spray for burning in the combustion chamber. When we tackled the Neptune engine, it started life with a scaled-down aero-system and a spill burner, as described above. The fuel pumps were supplied by Plessey and the other controls by Lucas, both mainly concerned with aero parts. This made it a very expensive and complicated unit. Joe Poole (development) had his turn at our battles with the Plessey-Lucas fuel system for the aircraft auxiliary power plant (AAPP) for the Avro Vulcan bomber, which plagued our lives at that time through its unreliability. (That story appears in Chapter Six.) Joe recalls one day when he started at 8.30am as usual and didn't go home again till after 6.00pm the following day. They needed - and demanded - from us a lot of very patient and meticulous work. Joe had his own saying of that time: "Every improvement created two more problems!"

So, rather in desperation, we evolved our own fuel controls and combustion systems. We 'invented' a small three-piston swashplate pump with a very simple leaf spring governor, coupled to a commercially available simplex atomiser – the Demon 1. The leaf spring governor was attractive in its simplicity; just holding shut a half-ball orifice till the centrifugal force from the weight of the half-ball lifted it off the fixed orifice, fuel was bled, and engine speed was governed. There were a few cadmium-plated steel springs for trials but they broke due to hydrogen embrittlement of the steel, so we went to beryllium-copper and never had any further problems. This combination, with a simple adaptation of the engine exhaust temperature controller, worked excellently, being reasonably priced, under our own control and totally reliable. Some units reached several thousands of hours' life without trouble and governing normally achieved 1½-2% control but, when we tried to improve that to better than 1%, we ran into instability.

The only 'peculiarity' was the addition of a fuel accumulator to aid starting. Because the off-the-shelf atomiser gave too poor a spray to ensure light-up at low speed when the engine was started, we developed the practice of cranking the engine a few turns with the sprayer on/off tap at the off position. When the fuel pressure had built up sufficiently in the accumulator, the tap was turned on and a good spray ensured quick ignition and helped pulling away. This accumulator took various forms and the first was a small Pyrex tube, sealed at both ends and screwed into the fuel system pressure pipe. As the fuel compressed the air in the Pyrex, we could see the level rising as the air was compressed and opened the tap at the right moment. Because that bit of air then disappeared into the engine fuel, we had to unscrew and empty the Pyrex tube before each start – and then repeat the procedure. It sounds terribly crude and that is probably why it worked so well and was very cheap. For later fuel systems, where starting might be remote from the engine, we replaced the Pyrex tube by a Hycar Ball. This was a squash ball made of Hycar rubber, which was impervious to fuel, and the accumulator air was that inside the ball. That also worked fine but needed an on/off tap operated by the rising fuel pressure.

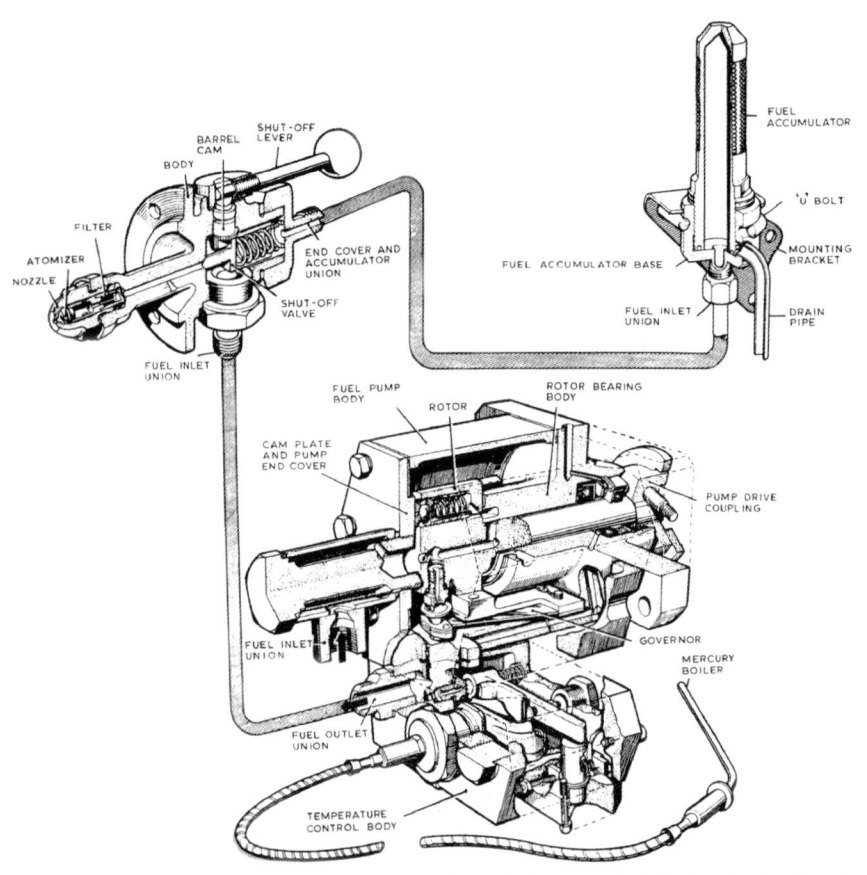

A cutaway diagram of the basic elements of the fuel system illustrating the leaf-spring governor in the pump. (IMechE)

Another feature we developed was a variable speed governor for engine speed control, where a second leaf-spring governor would be mounted on the fuel pump, set to operate at idling speed, so the engine would accelerate normally to that lower speed, probably around 50%. Then, via a little quill shaft mounted in 'O' rings, (fitted to be in compression, or they would wear), the vehicle's foot throttle would feed hydraulic pressure to a small bobbin, which tried to close that governor. If you pushed the foot throttle hard, this governor would fully close and the fuel system would enable the engine to accelerate to the point where the normal 100% speed governor took over. In this way the engine was stably controlled from 50% to 100% speed by the will of the driver. We also made further developments of this system where alternative 'instructions' were fed to the quill shaft and bobbin or, in a few cases, where a fuel-trimming governor was connected to the power turbine section of the engine. The system was very adaptable and gave very little trouble.

Another part of our 'simplification' came in the ignition system where the auto unit involved a magneto and large capacitance and circuitry, and was also expensive. The Project Department contrived, with the help of a consultant, Frank Ramsay, a very much cheaper, high-energy spark unit. This operated through a normal high-tension car spark plug. As it only sparked for a few seconds at a time for each engine start, we reasoned that the expected limited life for the cheap spark plug under this big-kick, would not be serious – and so it proved.

Earlier, I mentioned the 'cheap and nasty' Demon 1 atomiser which, I believe, cost only 10s 6d (52p) but we later became aware of a device called an Air Assisted Atomiser. In this, a second swirler is placed outside the fuel swirler and, when a small amount of air is fed to this outer unit, whatever fuel is injected will come out as a very fine conical spray – exactly as the combustion chamber wants. It was easy to manufacture and all our vehicle engines, and some production engines, used it. With the much-improved quality of the fuel spray, reliable ignition could often be achieved with a simple high-tension spark, avoiding the more expensive high-energy system referred to.

The piston pump had a stationary thrust face on which the pistons in their body pushed, with small ports to each piston. We devised a very effective 'weak extinction control'. We did not want our flame to go out and cause the clouds of noxious white vapour described earlier. This was accomplished by dividing the delivery annulus into two: one large and one small. When operating, the two feeds were externally connected and the pump and governors worked normally. The clever bit was that the governors could only bleed fuel from the larger annulus, while the smaller annulus kept feeding the burner, whatever the big annulus gave (which would be zero when the low speed governor was bleeding off). This small flow was sufficient to keep the flame alight, although the turbine was slowing down. To adjust this small flow to suit the engine combustion and application, we allowed the pump end cover, carrying the camplate, to rotate a little before clamping firmly to prevent movement and leakage. All automotive units, plus other variable speed engines, fitted this control and it was totally reliable.

In 1954, Joe Poole's first turbine work was testing the 1S/60 on 'towns gas' and our development of an appropriate control system. We started with six pressurised bottles, which worked well, but we had to change these so frequently that we decided to connect up to our local gas main in the factory. The first time we ran, we put all the (local) Rover paint shops out of action and were very unpopular! After that, we could only run after 7pm in the evening. At a later date, we tried to operate on methane, which had much lower calorific value, and experienced difficulty in reliable lighting-up because of the tight control needed on the air-fuel ratio. Adapting the fuel system was done by using the fuel pressures of a dummy kerosene fuel system to alter a large gas control valve, which had further adjustments to allow alteration according to the fuel's calorific value.

This market became attractive in 1965, when North Sea gas was first being tapped, and became high profile.

By now Joe Poole had relinquished the fuel system development to Tony Martin (Development), but it seems as though the left hand did not know what the right hand had done. More specifically, Tony found he needed to invent a new tube sprayer to inject the gas into the combustion chamber, beginning trials at a Midlands sewage farm. When it was working reasonably, he went out to Easington – a bleak, barren beach in north-east England, where large gas pipes came inland from the North Sea. In winter, this job was cold, cold, COLD! BP was our first contact and their Easington site was really "just a great muddy puddle", as Tony Martin described it. He and Don Proctor baled out the puddle with buckets and got a pipe blended into the huge pipe coming in from the North Sea. This pipe fed our hand-start engine. Knowing that the calorific value of this gas was higher than the methane on which they had set the gas fuel system, the control system was set as low as possible. However, when the engine finally lit, and pulled away, it screamed off like a mad thing! Much to their relief, the governor held and the engine settled down normally. We used to take some risks in those days!

The next development was quite exciting for us. We devised altitude-controlled camplate rotation for our piston pump by mounting the camplate in a bearing, and rotating it by bellows which sensed the absolute pressure around the engine – like a barometer. If the engine was operating at a high altitude, the camplate was well rotated and this reduced the acceleration flows considerably, without affecting the governors mounted in the pump. When testing at altitude conditions, we fitted a TV camera to watch the camplate rotation as altitude was increased. This system, coupled to our air-assisted sprayer, worked so well that we licensed our patent to Lucas, who made and sold back to us aero-standard quality fuel pumps with this built-in altitude compensation.

The combustion chamber also started as a scaled-down aero engine unit and was desperately expensive – as well as not being particularly brilliant for temperature profile into the turbine. Here again, we designed a chamber from first principles, where the primary part burnt a small proportion of the incoming air at near the necessary stoichiometric air/fuel ratio. Then the intense flame (2,000°C plus) was cooled by the rest of the inlet air to bring all to around 900°C for the turbine. Because of the shape required to do this effectively, the chamber was known as 'The Pepperpot'. In various sizes, this type of chamber persevered throughout Rover turbine work, except for a compact annular chamber used for certain applications.

To help the development, we had two water-flow models. The simpler model examined the main reverse-flow pattern in the primary zone to ensure we had strong vortex control for flame stability. The other model was a full-size perspex chamber, complete with all primary air, dilution air, and skin-cooling holes through which water was pumped, loaded with polystyrene balls of similar density to water. A thin slit-light then traversed the chamber, giving an excellent three-dimensional analysis of the flow patterns, until we were happy with the result. Happy memories!

You should have got the general message of this chapter – which is that if something could be designed and manufactured to do the job, Rover did it. We certainly turned the gas turbine business upside down to explore how simple and cheap the engine could become and, obviously, it is very sad that, in spite of a lot of successes, we did not turn the original Wilks dream into reality. But then, no-one else in this world has – yet. Interestingly, the last chapter will show that the dream is not dead, but very much alive.

Pepperpot combustion chamber

CHAPTER FOUR

T3 and T4, and twin-shaft engines

Back to the Wilks' dream of a production car gas turbine. The T8 engine and JET1 had been a wonderful experience and stimulation but were not very practical in realising the dream. So the Neptune and Aurora continued the chase. The Neptune went on to become the 1S/60 (our exploits with that engine are described in Chapter Five) and the Aurora or 2S/100 had Rover's first heat exchanger, the secondary surface recuperator. Rover's next try for a turbine car was again by taking a P4 car but this time mounting an Aurora in front, driving the rear wheels, as in a normal car. The exhaust, however, was ducted to the rear through a hollow chassis and to say it was unpleasantly hot would be an understatement. Noel Penny recalls travelling in this car, finding the seats almost too hot to touch and thinking it would be marvellous for crossing the North Pole.

After these attempts, another P4, MAC273 - really our T2 although it never got called that - had its Aurora engine fitted behind the rear seat with the exhaust going up a chimney behind the split rear window. From 1952, this car formed a practical and mobile test bed. The engine Aurora was giving 100 horsepower and working up to 3.8 : 1 compression ratio at 52,800 rpm. With so much weight on the rear, it steered like a bus (Tony Worster's expression – and he said Spen refused to drive it). Partly no doubt to help our funds, the good and faithful Admiralty Laboratory bought some of these Auroras, but without heat exchangers.

In 1950, Rover built a bearing rig, blown round by a small air turbine, to compare performance between different types of bearing (deep groove, annular contact, inner or outer located cage, roller, taper roller) and the offerings of different manufacturers. Because this rig made rather a penetrating high frequency racket (running at up to 60,000 rpm), Jack Belfit made a large sound-proof cover, which had to be raised for testing the torque loadings but otherwise was kept shut. He devised a cut out if the torque rose excessively. Jack also was quite a 'card' and seemed to get involved in all sorts of strange stories and events.

For example, Jack had written out some verses and when I came in and read them, I asked Jack if he had really written them as they weren't bad. Jack answered, "Yes, I wrote them but they were composed by Rudyard Kipling!" At this point I collapsed in laughter.

The MAC 273 car with its prominent rear exhaust

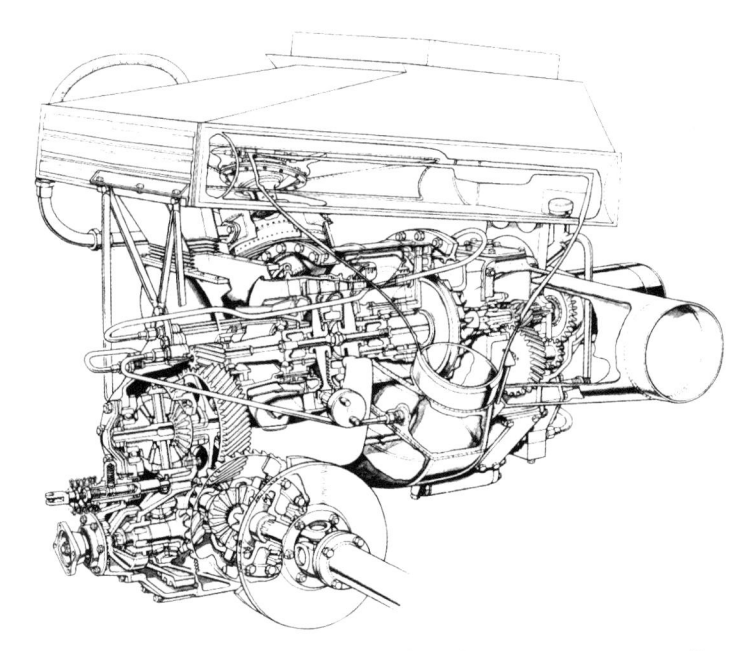

A cutaway diagram of the Aurora engine and gearbox   *(Autocar)*

Spen King decided that a 'proper' car should be created, with four-wheel drive. It should be small to reduce weight, of low-drag resistance and, of course, really attractive. Hence T3. Designed in the space of 12 months, it was a remarkable effort – really a big 'foreigner' but authorised by Rover's own management. This car was shown to the world at London's Motor Show at Earl's Court in 1956, presented as a Gran Turisimo two-seater. The car had disc brakes all round and four-wheel drive with a De Dion rear axle. The rear-mounted engine drove the rear wheels through a differential and the front wheels through a free-wheel and another differential – permitting them to over-run the rear wheels on cornering.

The light blue fibreglass two-seater was effectively designed by Spen, with Gordon Bashford, in the evenings at Gordon's home. Spen had just had an argument with snow and skis and had broken his leg. We have no doubt the design work did wonders to augment the natural healing process of his limb. Ron Hill remembers how a huge board was set up, all down one wall of our drawing office and a full-size design drawn there. Spen first had a slave prototype, called the Base Unit, built with no body and got the suspension, steering, etc. well sorted on this car before embarking on T3 proper.

Quoting H B Light again:

"This little glass-fibre bodied car would accelerate from 0-60 mph in 10½ seconds and from 0-80 mph in 18 secs. It had a fuel consumption better than 14 mpg at 60 mph.

On the high-speed test track at MIRA, a lap of 102 mph was timed, with plenty of power in hand. Maximum speed was 115 mph.

The Rover T3 car was of course a preliminary design, making full use of the advantages of a turbine-engined car, such as lightness of the unit relative to the power it can develop, absence of radiator or other cooling equipment, clutch and multi-speed gearbox. The car was a roomy two-seater saloon of small overall dimensions and lightweight. With the engine mounted at the rear, it had been possible to design a body having a low bonnet line which, together with a deep wrap-round windscreen and large rear window, gave exceptionally safe visibility. Four-wheel drive, too, was also considered a desirable safety factor on a car that had such a high torque to weight ratio.

The 2S/100 engine was a development of the well-known 1S/60 industrial gas turbine and consisted of a single stage centrifugal compressor with a maximum speed of 52,000 rpm, driven by a single stage axial turbine, re-designed so that it took only sufficient power from the gas stream to drive the compressor and fuel and oil pumps. A second single stage power turbine was added to take the remaining power from the gas stream and to drive the front and rear differential units. This reduction gear also incorporated a reverse gear, which could be selected by a central control lever.

The plate-type secondary surface contraflow heat exchanger, mounted on top of the engine, took heat from the exhaust gases to heat the compressed air before it entered the combustion chamber. The exhaust was ducted at about 200°C to a square opening in the top of the boot lid, which also incorporated an ejector orifice to ventilate the engine compartment. At 52,000 compressor rpm, the engine developed 110 bhp, with a pressure ratio of 3.85 : 1, a maximum temperature of 830°C and an air mass flow of 2 lb/sec. The self-sustaining speed of the engine was 15,000 rpm.

Whilst the heat exchanger fitted to the 2S/100 had brought about a marked improvement in fuel economy, it was evident that its life was not satisfactory. Frequent checks were carried out on the power unit in the car in order to obtain a measure of performance reliability. Two interesting and import points emerged:

1. The heat exchanger gas side pressure drop became almost prohibitive in approximately 3,000 miles. This was due to fine carbon deposition in the matrix of the heat exchanger on the gas side.

2. Although the Company's industrial experience had indicated that fall-off in power could be quite catastrophic when no air cleaners were used, this was not evident from the running carried out in T3.

As with all turbo cars, the only pedal in addition to the accelerator is the brake which, together with the handbrake and the reverse gear, constitute the total controls. The four instruments under the eye of the driver were for jet-pipe temperature, compressor rpm, speedometer and combined oil pressure, fuel level and ammeter."

Because the transmission was simple with no synchromesh, manually shifting gear from forward to reverse, or back, had to be done with the engine stationary or idling. If the action was not firm and swift, it was possible that no engagement would take place and the engine would have to be shut down to prevent the power turbine running free. A little traumatic – but it was never a serious issue.

However, shellacking of the gas side matrix concerned us so, in 1958, T3 commenced an organised endurance programme to examine the rate of fouling in the heat exchanger as miles built up, both on kerosene and on diesel fuels. The car was pretty busy as a demonstrator, often called up by Rover so, while doing this work, great care had to be taken to keep the car in 'easy to restore' state, with the extra instrumentation quickly removable. T3 covered 5,000 miles in these exploits at about 7 mpg and was very nice to drive. Tony Worster, his foreman Bert Hole, and Peter Candy worked out a number of test routes, to Cambridge, round Cotswolds, Fosse Way, Motorway M1, and London, driving in pairs. Every ten hours they called at MIRA at Lindley to record engine data. As Peter (a test fitter who became a Development Engineer) describes it, "you read the manometer on the straight and the pressure gauges as you went round the banking". Quite exciting!

The 2S/100-powered T3 car, with four-wheel drive: the 100 bhp engine is positioned over the rear axle.

Another hazard was Thursday lunchtime at MIRA, where every week lunch included spotted dick (steamed suet pudding with currants), which lay rather heavily on the stomach of the driver, although the car didn't seem to mind.

Peter also recalls one occasion when a high-pressure nylon pipe caught fire. I had introduced this as being easy to install, flexible, etc. but it had been routed too near the inboard disc brakes. Peter had just descended the steep hill to Northleach on the Fosse Way and, as he stopped for lunch, saw smoke from the engine compartment and had to do some prompt extinguishing.

On a different tack Tony once drove T3 back from London in thick snow and was pleasantly surprised how safe and quick the journey was in spite of the lack of feel on the throttle.

Fouling of the heat exchanger, caused by shellacking of carbon on the gas side of the heat exchanger, happened slowly with kerosene, but rapidly on diesel fuel. We never found a really practical way of removing the shellac except by immersing the whole heat exchanger in a hot trichlorethylene bath. We developed a 'black spot' rating system to calibrate the carbon precipitation quality of the combustion system – this drew mixture from the engine's exhaust, passing it through a white porous paper for a fixed time. The black spot was then calibrated by a simple light meter. It was an interesting system and we noted the effects of more or less air assistance, hotter or cooler primary zones in the combustion chamber, as well as the effect of different fuels. At one point I attempted a rather drastic cleaning trial, where water was fed into the exhaust of the engine, hopefully to 'shock-off' the shellac. The only effect was to cause many more irreparable cracks in the stainless steel structure, plus one rather red face! (For an interesting follow-up on this problem, see Rolls-Royce action in Chapter Eleven).

By this time, a new ground-breaking prototype, the T4, was getting the attention and T3, although remaining as a demonstrator for a little longer, was put on one side. It is now in the Rover Heritage motor museum at Gaydon as a static exhibit, as its engine cannot run.

Whilst this vehicular activity was taking place, the turbine department had, from 1957, been designing and developing a new gas generator, partly to reduce cost and also to make less demands on oil and bearings caused by high temperature heat soak. This gas generator used our first cast, radial-flow turbine, as was becoming popular for turbo-chargers in trucks and cars, and was of much lower cost than the previous all-machined turbine. The new design took the bearings away from the heat by overhanging the compressor and turbine components. To avoid an inevitable whirl problem, the roller bearing outer track was mounted in a springy assembly with its movement damped with a thin oil squeeze film. The whirling speed of around 8,000 rpm was passed through on the electric starter with no ill effects noted, and thereafter the assembly ran smoothly from 35,000 rpm idle to 65,000 rpm maximum power. Effectively, the gas generator ran on its mass-centre and this made it very smooth and mechanically quiet. If the squeeze film oil supply ever failed, the engine simply would not start, only graunch around at 8,000 rpm, though this virtually never happened. Later, we also did experiments of having a firm rubber insert around the roller bearing, replacing the rather long and expensive spring mechanism, but still using oil damping. That worked fine and was adopted for a Thrust Jet we made later (see Chapter Nine).

This overhung assembly was the basis of our next engine, the 2S/140, and we built a one-off Adaptation engine with this type of gas generator. It was fitted to the T3 Base Unit and gained useful experience. Later, this same engine was fitted with one bank of our first primary surface recuperator as already described.

Later still, the same long-suffering T3 base unit was fitted with a ceramic heat exchanger in the exhaust to see if it would tolerate being hot and bumped around under car conditions. We took it over the pavé test area at MIRA and the ceramic passed those tests with flying colours.

The 2S/140 was the culmination of all Rover's turbine experience. The unit combined many novel features, such as the primary surface heat exchanger, a cast gas generator turbine, variable-geometry swirl vanes into the compressor, and movable compressor diffuser blades to achieve a low idling fuel consumption. This latter, called 'half-flow idle', was quite an ingenious device which operated the diffuser blades via compressor delivery pressure acting on a piston, sealed by a rolling diaphragm, effectively reducing the air flow and 'size' of the engine at idling speed. However, the fuel economy benefit was small.

The main casings of the engine were of sheet metal and mounted between the heat exchangers and much expense was reduced by using a good quality mild steel, called Corten, insulated with segments of moulded insulation. Rover decided to combine all the benefits from our turbine research, and that of our suppliers, into a car with the 2S/140 engine to demonstrate progress to the world, and to explore also our latest attempt at a feasible gas-turbine-powered production car. They also took advantage of the car, T4, to 'try out' a design considerably different from any previous Rover model and see how the world reacted before they launched the planned P6, known generally as the Rover 2000, subsequently Rover's best selling model ever when it appeared in 1962.

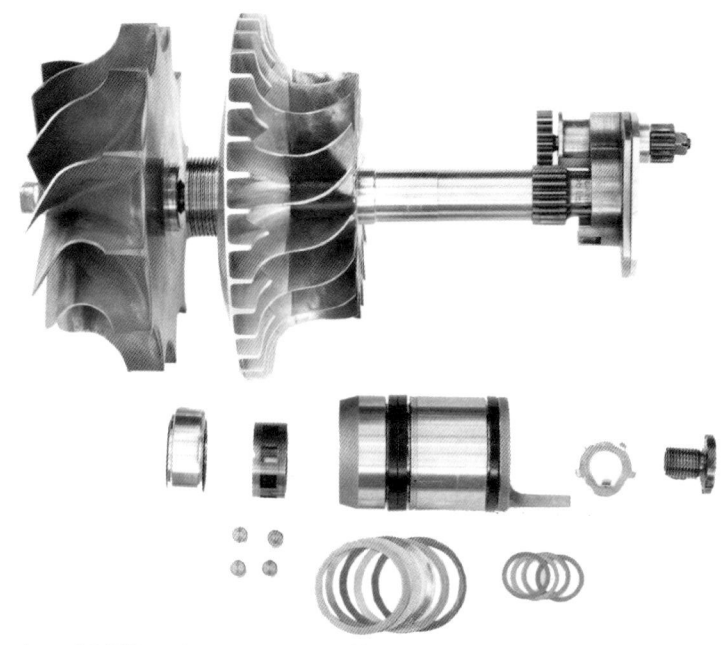

The overhung 2S/150 cast gas generator turbine with mounting spring.

T4 came in 1961 and was the seventeenth prototype 2000, but with a longer nose (approximately 18 inch) to accommodate our new turbine engine. She was prepared, as usual at the last minute, for the Motor Show and Martin Hurst, Rover's latest Managing Director, came to see her at 8pm on the evening before delivery to London. He thought the two instrument binnacles on the dashboard should be changed round, so we all agreed – "What a good idea" – and left them alone!

Quoting from H B Light :

"T4 was an attractive, conventional-looking, four to five seater, four door saloon, having an advanced specification including front-wheel drive, disc brakes on all wheels and fully independent suspension.

The 2S/140 engine forms a complete power pack incorporating the full transmission, front wheel braking system and all auxiliaries. The two paper-drum intake cleaners are located behind the headlamps.

The engine exhaust is taken via a bifurcated duct attached to the engine at one end and at its other end to a single duct running along a tunnel over the rear sub-frame to two wide, shallow ducts at the rear of the car. Considerable technical advances had been made with the engine design and compactness, particularly in regard to improved fuel consumption."

The 2S/140 twin-shaft 140 bhp engine with recuperative heat exchanger, as installed in the T4 saloon car.

To avoid the slightly tricky forward/reverse gear change of T3, T4 had a cone clutch operated by engine compressor air, but it was marginal. Too little air pressure risked slip and too much caused the cone to jam, so that gear change turned out considerably more of an embarrassment than with T3. At a later stage, Pete Candy and I replaced the cone clutch by a simple dog clutch, operated by the air pressure. It worked well, though Spen wasn't too pleased. To improve low speed torque, a two-speed automatic transmission was tried but was not deemed acceptable. About this time, work in Rover Engineering on the V8 powered prototype, known as the Three Thousand Five, was intense and Pete Candy was asked once at MIRA to take a prototype, fitted with a 3-litre 6-cylinder engine, back to Rover. On the way, going over a hump-backed bridge, the car slid off the road and Pete rang Spen to tell him of the mishap:

Spen: "You were going too fast"
Pete: "No, I wasn't"
Spen: "You were going too fast"
Pete: "No, I wasn't"
Spen: "You were going too fast"
Pete: "Yes, Spen" – and that was that!

Pre-production Rover T4 Saloon GT, 1961, which was the third public application of a Rover gas turbine car. The engine is mounted below the bonnet and drive taken through the front wheels.

T4 was introduced at the Motor Show 1961, before announcing the new Rover 2000, and Rover was pleased with encouraging public reaction to the body shape. There was an intensive test schedule planned for T4 but, because other car manufacturers, notably Chrysler in the USA, were also pressing ahead with turbine research, it was decided to interrupt our development programme to show the car at the prestige exhibition at the New York Show in April 1962.

Again quoting H B Light:

"Prior to being loaded into a Seaboard Canadian CL-44 cargo plane for its flight to New York, the Company's latest gas turbine prototype car was demonstrated at London Airport before pressmen and photographers. Accompanying T4, was the world first gas turbine car, JET1 of 1950, which was also to be on exhibition at the New York Motor Show."

In New York, George Heubner of Chrysler met Spen King and they drove one another's turbine cars on the New Jersey turnpike. George was suspicious of T4's road characteristics and tried, but failed, to make the front-end dig in as he braked hard in a turn. T4 actually had good handling characteristics whilst Spen found the Chrysler ponderous, underpowered and uninspiring. Rather typical of contemporary US motor cars. (For an independent assessment of the Chrysler car see Appendix F).

Further from H B Light:

> *"After creating much interest in the US, T4 was returned to the UK for a continuance of tests and for further development. In June 1962, however, T4 was in France at Le Mans where, preceded by 3 outriders, it was driven round the 8-mile course before the 24-hour Race amid applause of the spectators.*
>
> *During the press previews in September, prior to the announcement of the 1963 Rover Car range, members of the press were once more given the opportunity of seeing the Gas Turbine car and, this time, were able to ride in it.*
>
> *Back in France again at the beginning of October, T4 was shown and demonstrated to the French Press. This was part of the New Car Announcement Programme at the Paris Polo Club, just before the opening of the Paris Motor Show.*
>
> *In between these public appearances, intensive development work continued on T4 at Solihull."*

T4 was the only turbine car we built which always had heat exchangers fitted. The high-speed rotating components of the 2S/140 were also made up into an attractive, non-heat exchanger engine, the 2S/150 with a Rover Pepperpot combustion chamber. This unit was used for a number of applications, as recounted later.

The general performance of the car was acceptable with a maximum speed of 115 mph and 0-60 mph in about 11 seconds, but the fuel consumption was disappointing at around 12/14 mpg generally and a best of 19 mpg. Crucially, the potential production cost of the engine was too high, reflecting the amount of high temperature alloys still employed, with no significant potential for these to be reduced. The only glint on the horizon was the (then very theoretical) potential of ceramics to replace these expensive elements (eg nickel, chromium, cobalt) but Rover decided, at this stage, that there seemed to be little hope for the Wilks' original dream to be realised. The Wilks had, by this time, also relinquished the reins of the Company – we had become part of British Leyland – so the initial motivation was gone.

Chris Bramley (development) left turbines during the programme on this engine, to start working for Rover Research on car programmes, but was soon allocated to do strain-gauge jobs for us, so his escape was brief. Apparently, we asked him to gauge up running rotors at speed and temperature – at a time when this was pioneering technology. Chris experimented with ceramic and pyrocements to 'stick on' the wire strain gauges – about a quarter-inch square – on the turbine blades, taking the leads through holes machined in the turbine blade platform, down the disc, along the shaft to sliprings working at up to 50,000 rpm. Heady stuff.

Into this rather gloomy world came an American glass ceramics manufacturer – Corning Glass. Corning was developing a glass ceramic material, out of which they had made a matrix very similar to the metal regenerators used by the American turbine researchers at Chrysler, Ford and Detroit Diesel (GM). Because these metal discs distorted, due to the temperature difference across the disc, the task for the seals was onerous and always the 'Achilles heel' of the system. The good news, however, was that it was possible to achieve over 85% thermal efficiency.

The potential advantages of the Corning ceramic material were: very nearly nil coefficient of expansion; very low thermal conduction and a cost expectation of under £10 eventually in production. Though we had achieved 80% thermal efficiency from the primary surface recuperator, we still had serious life limitations on metal cracking, carbon deposition and very high cost, we felt we should have a serious try with the Corning ceramic.

We rapidly redesigned the new 2S/150 engine to accommodate two 18-inch diameter Corning discs; one on each side of the engine, with drive units of big diameter gears mounted on each disc via a spring, to ensure the ceramic was always held in compression. We even tried a single long leaf spring, which wrapped round the disc and that seemed to work well, too. Drive was at around 20 rpm at full power. Sealing the high pressure was effected by a seal surface, sprayed with nickel oxide and mounted on twin bellows by which we could easily alter the loading. Later we developed a cheaper leaf spring design.

On the casings, however, as Joe Poole described it – and he was Chief Designer then – the first main casing was a disaster. As soon as we pressure-tested, it blew up in all sorts of strange shapes and was quite unusable. A quick redesign, with a large perforated tube sheet from front to back, worked fine.

We used to have and try out good ideas, like these:-

Because the nickel-based alloys we used to make turbine shroud-rings expanded more than the turbine and lost efficiency, we decided to try out alternative materials, which would grow less. These alloys clearly behaved correctly in the laboratories to produce their desirable characteristics but, in a real engine, they were not as successful. They expanded less but also distorted such that we suffered turbine tip-rubs (just like in the early days). After several trials, we pronounced them unsuitable and 'piggish'.

Another idea to shorten the gas generator-shaft, and its overhang, was to use a Sealol face seal to prevent lubrication oil entering the compressor. The first test was run, and, as the unit accelerated, it refused to respond to the throttle and was 'running-away' – a dangerous situation. Don Proctor (Chief Tester becoming Supervisor), who was at the controls, picked up a pair of overalls and stuffed them into the engine air intake. That stopped it! The engine was running away by using its own lubricating oil as fuel, and the earlier experience and instructions of fifteen years before had been forgotten. Incidentally, we got the face seal to work well after that.

T4 engine compartment fitted with the 2S/140 engine

CHAPTER FIVE

Rover Gas Turbines Ltd

The interest of the British Navy in our turbine efforts has already been mentioned and also their desire to encourage us by buying some T8s and awarding us specific development contracts. The Naval Research base at Admiralty Engineering Laboratories, West Drayton, supported combustion work on diesel fuel during the 1950s, as the Navy did not regularly use kerosene until jet aircraft embarked on carriers. The Naval Headquarters at Bath gave us our first orders for mobile water pump sets, to be adapted as on-board units on all Navy ships. Having introduced the matter of a research group in Rover supplying turbines for sale to the world outside, we now need to describe how this evolved.

Let us start with a person: Doug Llewellyn joined the turbine team in 1951 as a draughtsman under Ces Bedford in Project A. His first job was on T8 to re-route the air feed to pressurise the oil seals from the back of the impeller, thus reducing considerably the quantity of air used whilst still sealing satisfactorily. The result was that the engine then gave nearer 250 hp instead of 200 hp. That engine went on to be the unit used in Belgium for the speed record. He then helped design the Neptune engine, which was to become, in due course, the 1S/60. Whilst Ces Bedford designed the castings, mostly of LM8 aluminium anodised to improve resistance to corrosion, Doug dealt with the sheet-metal parts. He had to conform to the then current position of Rover and the Sheet-Metal Workers' Union.

Before the Second World War, Rover and this union had fallen out, and Rover management still refused to employ any member of that union with the result that, within the Company, panels could only be designed with single curvature. This explains the angular shape of the first Land-Rover and, in our case, the way our main casing was formed. In fact, Doug drew out the developed blanks to assist the tool room to make them. Then he tackled the oil and fuel pumps. The original oil pump remained throughout the production but the fuel pump was replaced, first by a small Plessey pump, then by the iniquitous 'shrunk-aero' unit and, finally, by our own piston pump.

Now let us look at the most interesting item to make and assemble – the turbine shaft, fitted with the compressor. This steel shaft had seven diameters machined, some to a 0.0004-inch tolerance and to a concentricity of 0.0002-inch. First the aluminium impeller was heated and shrunk on the shaft. Next, the steel rotating guide vane (RGV) had to be heated to its limit in boiling oil, its vanes lined up carefully with the impeller, and shrunk-fitted. Here we met trouble, as several units moved slightly during assembly, making the compressor inefficient and a worry for blade life. To overcome this the shaft was frozen to -30°C before being pressed into the RGV. So far, so good, though we were stressing the RGV hub past its elastic limit. Also, we discovered we were boiling the RGV in oil at a temperature which was at its flash point, so we had to change to a small electric furnace. Then we found some shafts were bending like bananas. Rover's Chief Metallurgist, Sidney Heslop, came to the rescue by discerning that the -30°C freezing of the shaft had exceeded the heat-treatment tolerance. Next we had to alter the shaft heat-treatment temperature to -40°C and, having done that, our high-speed shaft remained straight. However, we found one or two engines exhibited 'singing' RGVs - ie they were not touching the impellor firmly. So, for a while, we fitted a metal shroud around the outside of the RGV blades and copper-brazed the contact points, resulting in some loss of compressor efficiency. Later, we restored the taper nip between impeller and RGV blades by increasing the assembly pressure. Rolls-Royce had apparently encountered exactly the same problem on aero-engines in the 1940s – all the best people make the same mistakes.

The Turbine Team realised that the effort to support the attempt to go into production needed more help. Doug was joined by a new man, Fred Hulse, who added a lot of 'production thinking' to our work. Fred figures in another funny story: Near Christmas time, Fred asked Doug to pick him a fair-sized holly bush from the roadside and explained that, traditionally, his family always pulled this down the chimney on Christmas Day. When Doug asked if his wife protested, he replied that she did and that was why he did it – to show who was the 'Gaffer'!

RGT (Rover Gas Turbines Ltd) was formed in 1954 and Geoff Searle was appointed as Managing Director. Peter Wilks became Production Manager for a couple of years and Doug was appointed his Technical Assistant; Derek Aslin, the Service Manager; and George Cowan, the Sales Manager. A pre-fabricated Coseley building was erected next door to Project A and the RGT folk moved in. Peter Candy recalls how he did an apprenticeship at Rover and got himself placed in turbines, under Nobby Clarke (foreman) when the RGT group was forming. At that time, there were only nine people in the Coseley building of which six were apprentices, including development engineers Laurence 'Tod' Butler, Fred Court and Peter himself. Doug used all the facilities of Rover that he could to manufacture the 1S/60 parts, but found that RGT were always second priority to the remainder of the Company, so it was hard work.

They, of course, fell behind planned production delivery dates and, once again, our Admiralty friends helped out. They were awaiting delivery of six engines and accepted six large crates of 'engines', which actually contained bricks and stones! Later, of course, the crates were duly returned for 'overhaul' and the real engines delivered. Rover Tyseley machined many parts, including grinding the critical high-speed gear teeth using the Swiss Maag Grinder.

We will quietly admit that the first prototype 1S/60 built by RGT gave all of 12 hp at its first run. The first production 1S/60s (now up to power) were for a water pump for fighting fires; we received contracts from the Navy for this pump and they insisted that it should be independent of any outside power. Therefore, people-power was the only option. We first tried a foot start, with a man lying on his back on the ground and pedalling a chain drive. Fortunately, we soon changed to the regular starting-handle principle, which also turned a magneto for ignition. The Navy had what they asked for but it was pretty hard work. The specification finally changed to provide a double handle. The unit remained that way for many years and the Navy specified two water pumps on each ship, which amounted to a lot of units. In addition, the Commonwealth Navies followed suit and, at this time in 1964, the complete unit cost £1,264.

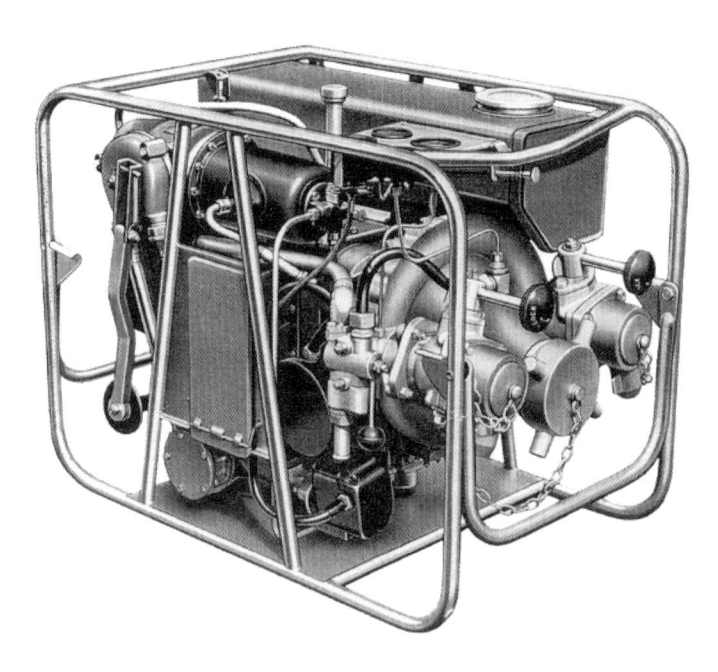

The 1S/60 salvage and general purpose fire fighting pump

Two more stories:

First - To impress visitors to RGT, Doug trained his pretty and diminutive Secretary to start an engine on her own. She was excellent but then needed to retire out of sight to recover while he got on with the negotiations.

Second – The Navy has a practice that *"If it moves, salute it! – If it doesn't, paint it!"* The 1S/60 water pumps were called 'Salvage and General Purpose Fire Fighting Pumps' and many of them had never been used before being returned to us for servicing, still complete with their on-board canvas covers. RGT had to hack off the much-painted and now solid cover, only to find the turbine unit seized with encrusted salt. After hosing them thoroughly, we usually found that they could still be started.

Altogether, 1,000 engines were sold to the Navies and they were much more popular than the heavy Enfield diesel pumps they replaced. Rumour has it that, to 'hurry up' the replacement, some of these diesels disappeared overboard - with no obvious cause.

Back on our ranch, RGT were exploring hard for new applications for the 1S/60 and one little incident deserves telling. Doug was doing a lot of travelling around and once arranged to use Peter Wilks' Rover 60. Peter asked him to check why the car had a poorer performance than it should. Doug knew that the Service Dept. usually retarded the ignition a little, and that the 60 was under-powered anyway, but he agreed to have a look. He found that the ignition was a little retarded, advanced it but noted little change. In playing with the throttle by hand, however, he thought it seemed rather unresponsive and discovered about four layers of thick Rover carpet under the throttle pedal, so it couldn't properly open the carburettor butterfly. Someone sometime had forgotten to cut off the excess bits of sound-deadening. The car was returned to Peter, who agreed it was now fine, but Doug didn't tell him for several years how he had cured the problem.

To the rather obvious question of sales of fire-pumps to civil authorities, the answer was always the same – the local fire brigades said it was up to the Home Office to recommend equipment and the Home Office said there had to be a specific request from one or more brigades. RGT never managed to overcome the Catch-22 situation. Someone did comment that the small portable fire pumps already in use were driven by a Coventry Climax overhead-cam engine and their exhaust was preferable to our big blast of gas. On one occasion, though, when talking to the Coventry City Fire Brigade who had just tackled an enormous fire in the Jaguar factory, where much of the plant and many cars were lost, they claimed that they really needed a 'fog machine'. An all-enveloping fog would deal with that kind of fire much better but they did not have anything like that. RGT decided to explore what they could do with a large-scale air-assisted sprayer, fed with the bleed air from our compressor and with water where the fuel would normally go. An experiment outside the Coseley Building was a great success. Solihull did not understand why they suffered fog that afternoon, when no-one else reported it. All our gas turbine engines on test stopped because the dense water-fog clogged their Purolator air filters and the excessive pressure drops cut the engines. Back to the Fire Chief, who then said that he had no money – end of story. Nevertheless, the fire-fighting unit was always an attractive possibility for gas turbines and an early showing of our water-pump was at Llandudno at a meeting for Fire Chiefs. Phil Phillips, who had been asked to write a technical paper about our engine, was there with one of these units, accompanied by Tony Worster. They put the unit on the beach and checked that it worked, went into the hall and gave the lecture and, amidst considerable public interest back on the beach, started our engine and fired water jets out to sea.

Next, the Army wanted their turn. Their need was for a fuel pump set which could perform two jobs. They wanted the 1S/60 to pump fuel from a tanker moored just off the beach, up to a large flexible fuel tank at the top of the cliff. When they were ready to move on, they needed the same engine to pump fuel from that tank to another several miles farther on. This was really two different requirements and Hayward-Tyler, who manufactured the chosen pump, asked for two speeds to permit the one pump to be able to do both jobs. Our standard output gave 4,500 rpm and that would do for one task. 3,600 rpm was needed for the other, so RGT designed a two-speed gearbox to satisfy these requirements. Land-Rover gears were found which satisfied, and coped with, this duty and gave RGT a useful alternative drive. When pumping fuel, our 1S/60 had to use the fuel being pumped which could be diesel, Avtur (kerosene or paraffin) or MT80 (petrol). Combustion was fine but the lack of lubricity of petrol meant that our fuel pump had to have silver plating of all rubbing parts made of aluminium-bronze. It turned out easier to standardise this feature on all our production pumps thereafter.

We had two 'advisors' during this project. One was the Head of MEXE, the Army's experimental establishment at Christchurch, and the other, Georges Roesch, from a London office. The latter had been quite a famous man in his own right, as Chief Engineer of Talbot Cars when they developed a splendid range of cars, many of which raced successfully in the 1930s. Not surprisingly, his suggestions were good and RGT were pleased to accept his advice. However, we sold only four of these fuel pump units after all that! (Author's note: Quite why the Navy people were at West Drayton, London, and the Army folk were on the coast at Christchurch, I never did fathom!)

Gas Turbines were all-the-news items in the 1950s and RGT had the wisdom to offer the universities an Instruction Set, which comprised a 1S/60 coupled to a water-brake and an appropriate number of instruments to enable students to calculate basic data from observations on their own tests. These were priced at around £2,250 and the British universities, to begin with, appeared uninterested. So the first sale went to Delft University in Holland. Then, universities round the world started to snap them up and, at one stage, RGT had sold a 1S/60 into more countries than Rover had sold Land-Rovers. Sales reached nearly 200 units. Interestingly, Doug Llewellyn, now Production Manager RGT, seemed too often to get called to start up a new set for its first run. He would find that the RGT instructions had been carefully followed but no-one dared to press the button! He did, and often found that our fuel system suffered from a common disease – 's..t on the half-ball' of the speed governor which prevented starting. He would clean it and all then went well. It was dirt, usually bits of cadmium plating, from their new high-pressure fuel piping to our variable-speed valve which thus by-passed the inlet filter. Nobody wanted to squeeze in a strainer on the pump rotor local to the governor, so our disease remained uncured.

ROVER 1S/60
GAS TURBINE ENGINE

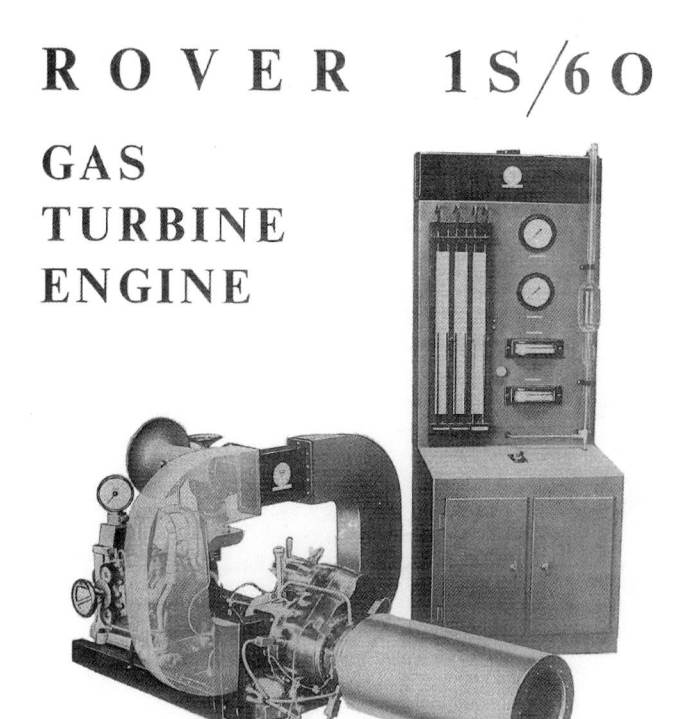

INSTRUCTIONAL SET
A Self-contained Test Bed Installation
For Universities and Technical Colleges

The IS/60 Instructional Set

Electric Generator Sets did pretty well, too. Our very simple leaf-spring governor gave acceptable speed control, without adding any feedback to improve limits. Petbow, Maudsley, ASEA, and Dale Electric all used the 1S/60 at some time and, in Chapter Seven, you will learn of generators sold to Vospers. Amongst other jobs, Dale tendered to the South African Government because they were considering a move – lock, stock and barrel – to Ascension Island if the pressure from the

African Black Movement became too great. They would then need easy and quick electrical power.

Another variant was the unit for Cloche Point on the Clyde in Scotland. A lighthouse there had been operating on two diesel engines, installed in 1905. One drove a generator for the light and the other drove a compressor to work an old foghorn. When one of them broke its crankshaft, RGT was invited to tender for a replacement, but it was obvious that the second engine could not be expected to last much longer, so we offered our engine with a generator to operate the light. We also proposed a relatively small amount of air bleed from it to work a new foghorn. This was enthusiastically accepted and, whilst being installed, it had to be set working quickly when a thick fog descended over the estuary. It was a Sunday morning and, after about an hour, the lighthouse keeper received an angry phone call from the vicar of Dunoon, across the estuary, protesting about the noise. It was so strident that his congregation claimed to be unable to hear his sermon. (It was, of course, the foghorn and not our engine!) A successful sale? Electrical generation involves a relatively heavy rotor, with considerable inertia. Starting the 1S/60 needed a centrifugal clutch between our unit and the rotor to permit our engine to reach near its running speed before we had enough power available to 'pick up' the rotor. Through Petbows, we got such a clutch from Twyflex Couplings Ltd, which started to grip at around 38,000 rpm and was fully 'in' well before governed speed. We used this clutch for a number of applications thereafter.

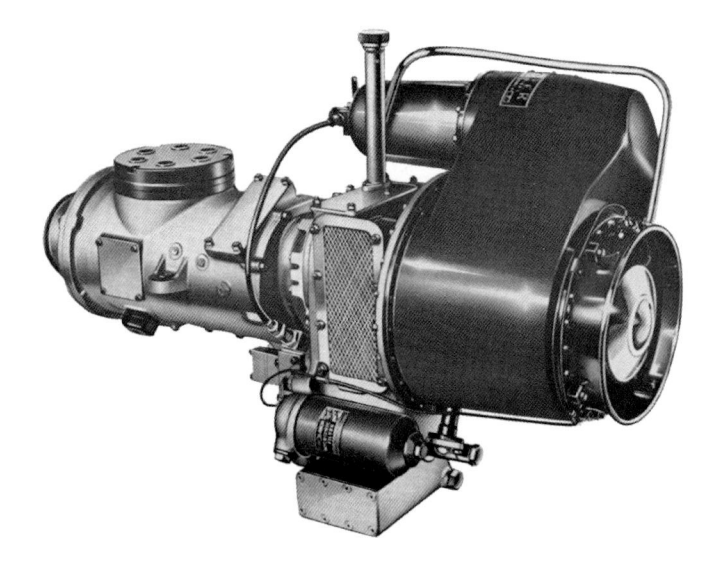

The IS/60 Generator Set, with BTH equipment, designed to produce 30 kVA at 400 c/s.

Tony Martin recalls efforts to improve engine starting, particularly in the hand-start units. We tried putting in fuel in excess of the norm at low engine speeds, then trimming that excess by a valve operated by P2 (compressor delivery pressure) to avoid compressor surge. He also remembers the relief when the combustion lit up and the engine pulled away during testing in the MIRA cold chamber at around -30ºC. Don Proctor describes a sand drier he got involved with, where the turbine exhaust made short work of drying up to eighteen tons of wet sand per hour. The 1S/60 was then sold to a market in South Africa.

One rather splendid customer for RGT in 1958/9 came from Canada. Catching fish was a pretty big deal in Newfoundland and generally they exported what they could and canned what they couldn't, but they still caught more fish than they knew what to do with. They had started to mush these excess fish, plus the bits leftover from the other processing, into animal food as there was so much protein in it. Then the 1S/60 came along and provided them with remarkable value in generating electrical power. This was particularly handy in isolated parts of the country for processing and drying the mushed-up bits of food for packaging – and it probably warmed their toes as well. All this was at about 70% thermal efficiency and they sold the food, called Fishmeal, at what was no doubt a handsome profit. Our units amassed huge lives, many surviving for several thousands of hours. That was the good news. The bad news (for us) was that they were a rough lot and the awfulness of what they did to keep our engines going beggars belief. Suffice it to say that when, eventually, we saw the engines again for overhaul, they came to us a heap of useless rubbish. Ah, well!

This was the time when 'total energy' was in the news and equipment was vital to provide 'no-break' for early computers, in the days when power cuts were the norm. RGT made a number of proposals, like that to one of the big banks in London – but little happened. One firm, called Braby, got interested in using the 1S/60 to operate a desalination plant – fresh water from sea water – but that also came to nought. Then, in 1971, natural gas became the rage in the UK and Chapter Three covered our efforts to be ready for that. Surprisingly, no sales followed.

When you offer an engine of specified power, inevitably you are asked for more. By 1956 RGT had extended their range with the 1S/90. This was the 1S/60 compressor diffuser with larger throats to increase the air throughput. They had also raised the turbine inlet temperature, using a better turbine rotor in Nimonic 105.

No-one now seems quite sure of the numbers but RGT had probably produced and sold some 2,500 engines by 1965. A number of us believe that, had there been a determined plan, Rover Gas Turbines could have become viable and become THE small gas turbine company of Europe. But it was not to be. The final moves of RGT are outlined in Chapter Ten.

CHAPTER SIX

Aeroplanes and auxiliary power units

In 1955, the Air Ministry gave Rover the first contracts to prepare the 1S/60 engine to become an Aircraft Auxiliary Power Plant (AAPP) for the Avro Vulcan Delta Bomber Mk II, which used electrical energy for its flying controls. Avro came to us and requested an AAPP to drive, at 8,000 rpm, a 400 cps 40kVA alternator, both for ground checks and to provide emergency flying controls. It became standard practice in Vulcans that our AAPP would be started before the plane landed so as to ensure power availability. This resulted in our engine becoming dirty, fouled by tyre rubber as our intake was near the plane's undercarriage. The prime requirement was that the AAPP could be started at altitude and give full electrical power within 10 seconds. This involved firing two 600 gram cartridges pointed at the turbine blades and, in the distinctly exciting two seconds whilst this happened, our engine would light up combustion, overspeed to about 51,000 rpm and settle on the governor at 48,000 rpm. There were another two cartridges in case the first pair failed to fire, and all four were cossetted in electrically-heated blankets. We also needed an oxygen feed near the igniter, to ensure safe light-up at altitude.

Aircraft auxiliary power plant for the Avro Vulcan, based on the IS/60 engine. It provided main engine starting and emergency electrical power.

Avro decided that, as their Olympus alternating-current starters were not completely reliable, our unit should also be capable of progressively starting their engines by air bled from our compressor. This also made the plane partly independent of ground services. For us, this entailed fitting movable diffuser vanes in the compressor to give enough air. As this AAPP had to be capable of starting, and operating, at 60,000 ft altitude and down to -50°C, we spent many weeks in the High Altitude Chamber at Lucas, Burnley – very cold and raw. We made about 116 of these units, at an initial price of £8,000.

At one stage we had to contend with metal slivers in the fuel system, which were eventually traced to dirty stainless-steel piping from a supplier. We found the cause and told Avro, who went berserk. The Olympus engines were fitted with the same manufacturer's piping – they got changed very quickly.

Because the Plessey-Lucas fuel system caused us so many headaches we were very relieved that our altitude-compensated piston pump was eventually adopted for all Vulcan units as a retrofit. I am sure that Vulcan crews must also have been pleased, if they had time to think about it.

The Vulcan, not only capable of carrying and dropping an atom bomb anywhere in the world, also had to cope with the even worse hydrogen bomb. This beast was so powerful that the only way it could be delivered was by 'throwing' it, at high speed from a height, in such a way that the crew had some chance of survival themselves. Unfortunately for him, our service engineer, Graham Woodhead – always called Timberbonks by us – was a passenger at the Farnborough Air Show when the pilot performed a demonstration involving a lot of negative 'g' loading. After the event and the aircraft had landed, Timberbonks reported at the RGT stand very white and shaken. A large piece of Vulcan wing panel was found to have rippled badly due to the high stressing and Avro were most displeased.

Avro Vulcan B2 bomber

Because Lucas were involved in the control system, they carried out some of the tests on a slave AAPP, in a little garage on a Midlands airfield. They had a well-instrumented van and a sharp technician, John Barnard (no relation) to help get the complicated system operative.

Another interesting comparison, the electric starter-motor for normal starting of the AAPP cost around £700, whereas our commercial unit came in at under £5.

There is a Vulcan at Wellesbourne, Warwickshire, used for demonstrations and Tony Martin, who used to be our trouble-shooter on the Argosy at Armstrong Whitworth, Bitteswell, had a good look round it recently, discovering that it still uses our AAPP to start its Olympus engines.

Our second AAPP project, in 1958, was for the Armstrong Whitworth Argosy troop and equipment carrier aircraft. Our engine drove a 28-volt 9-kW DC generator to operate the lights and a hydraulic pump to move the loading doors. It also charged the batteries used to start the Rolls-Royce Dart main engines. Being only a low altitude plane, we used our piston fuel-pump though, as already mentioned, we had to license Lucas to supply it. We sold about 100 of these engines.

The AAAP for the Armstrong Whitworth Argosy, providing electrical and hydraulic power for auxiliaries and main engine starting.

There is an amusing story about the Argosy unit's engine hours. Armstrongs expected our hours would be around a tenth of the Dart's but, in practice, we were surprised to find our AAPP actually ran more hours than the Dart. The reason was that there was a galley on the plane and our AAPP was constantly working hard to provide power for coffee, tea and other essential services.

An interesting project was the P1154, a Vertical Take-Off supersonic fighter, for which we made the main engine starter. Our 2S/150 aerodynamic components were built around a Lucas annular combustion chamber and thus became the 2S/150A, which was very compact and about 100 lb total weight. Unfortunately, the project was cancelled by the Government and the P1127, a subsonic aircraft, which became the Harrier, was pursued. But now our starter was too big, heavy and powerful. We were prevailed upon to direct a group of Rotax designers to draw up the 2S/75, a scaled-down 2S/150. They gave it a simple gear fuel pump with five flat-spray atomisers, made by a sawcut in a hypodermic pipe, with air emulsion for light up. A pintle, variable flow number, sprayer was also tried but rejected. The first of these 2S/75 engines was made in nine weeks and eventually there were over 2,000 sold by Rotax.

We also produced about one hundred and fifty 2S/150A AAPPs for the Avro 748, a small airliner powered by Rolls-Royce Dart engines. Our unit operated the starboard Dart's engine auxiliaries through a clutch to provide DC power for making the plane habitable (and warm) on the ground when the Darts were switched off. Joe Poole remembers an 'unfunny story' (now funny of course) about this unit. Eric Richter, in our Design Office, made an excellent job of the gear train, except there were two gears to be meshed and both had a left-hand helix. Last minute panic stations resulted and a new right-handed helix gear, with soft teeth because of the haste, was procured in days. Joe swears the first drawings were checked, and double-checked. Nobody was sacked.

The military equivalent of the 748 was called the Andover, and the Queen's Flight operated them. A four Tyne-engined freighter, The Belfast, also used our AAPP. I presume our fuel consumption and noise were much less than those of an idling Dart and also saved the Dart's hours between overhauls. We submitted for another AAPP comprising two 2S/150A units, coupled together to produce 300 hp, to be mounted in the VC10 airliner but nothing came of this plan.

The HS801 or Nimrod, based on the Comet IV, which used Spey engines rather than the original Avons, had fitted an ingenious AAPP, comprising a 2S/150A driving a 1S/60 impeller, directly on the power turbine shaft. This unit was mounted in the tail of the HS801 and fed air along a 50-foot long pipe down the plane to start the Speys. It also provided air conditioning for a mass of early warning equipment. Quite reasonably, all airborne units had to be very safe, and proven to be, both theoretically and by demonstration. For example, we did an unusual and interesting test on a 1S/60. We drilled some little holes near the blades/rim of a 1S/60 turbine rotor and then oversped it in an AAPP in an aircraft until something went bang.

When this happened pieces of hot turbine, which had been contained within the engine, then fell out of the exhaust onto the tarmac and started to melt the asphalt. Our unit was passed as safe. On another occasion, when we had to show how a failure would be contained within the engine, some tests were carried out in the Rover Project C hangar, well away from people and, whilst we were testing, an engineer from the next door hangar called us to tell us we were on fire. Apparently the protective stacks of old periodicals we put around the engine had been too close and had caught alight.

Because of the huge amount of energy contained in items moving at high speed – the rotating components of our turbine engine being a good example – we were always very sensitive as to how our components would fail, whatever the cause. Cast turbines, far cheaper than forged and machined rotors, therefore needed proving by being spun up to very high speed to (a) confirm it was a good casting and (b) see how the failure fragmented. Cast discs can break into two or three bits and the energy in these will be far more than we can expect to contain within an engine – ie pieces will come out very fast and could easily kill or maim someone.

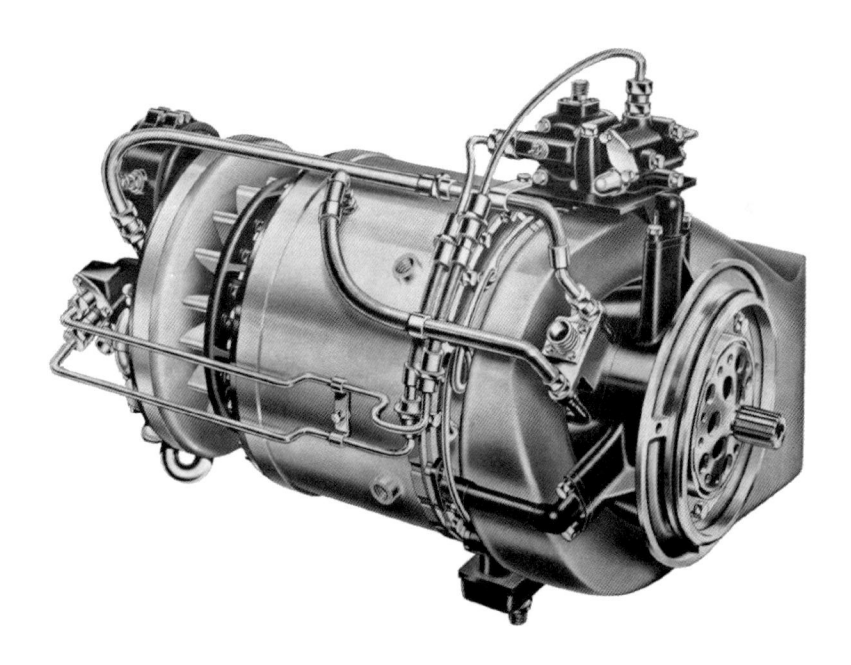

The 2S/150 main engine starter for the P1154 vertical take-off aircraft

The 2S/150A unit as used as AAPP on the Avro 748

The results of a 'successful' overspeed failure of an axial turbine

Some of the early cast materials we tried for turbines had good laboratory characteristics but proved to be brittle in failure and so could not be used. The way we proved them was to spin all castings in a spin rig. This worked off air pressure, which blew round a tiny air turbine mounted in a turbo-charger bearing assembly, driving a thin quill shaft – actually a knitting needle - with lots of damping washers along its length, and the rotor to be tested. This unit was mounted vertically and placed between two big and fairly heavy trolleys, carrying absorbent material (progressively: soft card, hard cardboard and wood) to catch the broken bits totally undamaged and ready for analysis. For many years, this rig worked excellently, though I will admit that just once (never again) we did not fasten the two protective trolleys together for a spin test. When the specimen broke and we were watching the rig through a window, we saw flashes and sparks around the test room and knew the broken bits were flying round very fast. So we got our heads down quick, waited and got away with that one!

When we found a successful material with enough ductility, we proof-spun every rotor to just below its elastic limit, probably about 20% faster than normal full speed. Provided the measurement we used as a check had not exceeded a specific degree of growth, we felt safe to use that rotor in an engine. This growth was usually indicated by the change of dimension of a critical piece of that rotor.

Tony Worster recalls that NGTE (Farnborough) took some very high-speed photographs of bursting rotors, too. We used to predict and record the overspeed behaviour till final failure occurred on all high-speed components. Items like the rotating guide vanes and impellor just grow so much that they lose the drive on the shaft and the engine stops – a nice safety feature.

Nimonic 90 axial turbine rotor

Nimocast 713LC radial turbine rotor

Turbine rotors are the tough ones. Ductile materials, such as Nimonic 90, grow so much that they invariably seize on the outside diameter – ie the blades in the surrounding shroud. More brittle materials, like Nimocast 713LC, which we regularly used, could be designed to fracture just below the blades. One or more of these blades, and their associated bits of rim, would then break off and could be expected to be contained within the engine. I can recall, though, that on two occasions, as we were developing the above procedure, we had turbine rotors burst in running engines – fortunately both on test beds. However, we never burst one on a car. We wondered how the cast radial flow turbine would cope and were grateful when a test showed the thin vanes 'unfolded' themselves at 95,000 rpm and rotation stopped again by blades rubbing on the shroud.

For one potential customer, we considered our engine for a Dessault NATO fighter, needing oxygen to run up to over 50,000 ft. The Burnley High Altitude Testbed (HAT) couldn't go 'high' enough so we designed and made an exducer, which should allow the HAT's turbines to give us that height condition. We did not get the contract, but perhaps we should have – for ingenuity.

The 1S/60 was a fine engine, but it was hardly likely to be the answer to a maiden's prayer and actually power an aeroplane, but it did – in 1959. One has to go back to an arch-villain again for this. There was a 'larger than life' individual named Vivian Bellamy, who flew aeroplanes of many kinds from the Hants and Sussex Aero Club, and he wanted to install a 1S/60 in a small plane. He was introduced to Geoff Searle and hence to Doug Llewellyn, who was told to have a go. Normal output speed of 4,500 rpm was too high; Viv said he wanted nearer half, so Fred Jones, in charge of the RGT drawing office, 'codged' up a gearbox by use of a Land Rover transfer box to produce 2,100 rpm at 100% turbine speed.

Viv Bellamy arrived at RGT in his V12 Rolls-Royce Phantom 3 (painted yellow and black) to collect the engine. The rear door was taken off the Phantom and Doug and Nobby Clarke somehow managed to squeeze the 1S/60 into the Rolls before Viv departed south again. He adapted a small biplane, called a Currie Wot, to fit the 1S/60 but it had only a 6-ft diameter propeller of far too coarse a fixed pitch. This would not allow the 1S/60 to start, because the temperature control stalled the engine. So Doug – still involved – tried bending the propeller (crudely forcing by jamming it in the hangar door) to reduce the pitch. This helped but was not enough, so Bellamy got out his own ex-wartime Spitfire, revved it up in front of the Wot to make a large draught and, between them, Viv and Doug got the Wot engine up to 100% engine speed. It rapidly took off, in about 20 yards, and went well – there was enough power but a smaller variable-pitch propeller was obviously called for. Back at RGT, Geoff Searle said, "Stop!" – but Doug was bitten by the bug. Dowty Rotol said they would make a propeller eventually, but too slowly, so Doug and Fred Jones designed a variable-pitch control unit (overtime without pay). Our 'foreigner' system made the bits, a friend loaned a propeller and a quick test round the Rover test track was carried out on the 1S/60 mounted on the back of a Land Rover.

The first run at night was fine so the RGT fitters, who had also worked for no pay, asked for a demonstration. Unfortunately, some rain was now falling and the propeller hit some sagging tree branches. After a quick trim of the propeller, Doug showed the system to Searle, who brought out George Farmer (then Chairman of Rover), Maurice Wilks and Bernard Smith. Probably not unimpressed, Wilks told Doug to put it in a plane, which he was delighted to do. Bellamy, of course, was also delighted and the Wot now flew magnificently from the Hants and Sussex Aero Club base at Eastleigh. The design of Doug's pitch control mechanism meant that power stayed constant, whatever the forward speed, and the Wot climbed strongly. Bellamy got reporters from Aeroplane and Flight, who both gave rave reports.

The next part of the plan was to show the plane to the Press at Elmdon – the old name for Birmingham Airport – but the Popular Flying Association insisted that the plane, and engine, must first fly for 50 hours. Bellamy obliged, getting very cold as the cockpit was open and the month January. Because the fuel tank capacity only allowed 1.5 hours flight, it meant a number of short flights and Doug was asked to do a refuel to save Bellamy some of the trouble. On the H & S Aero Club airfield, the kerosene tank was situated at a garage about one mile away from the runway and involved taxying round over a grass perimeter track. Doug found it quite an experience; dodging one helicopter on test, and ending up in a thick hedge with bits of propeller flying around, caused by a hidden concrete post. He was extricated and given a strong brandy. While recovering, he spotted a wooden propeller hung as decoration in the club, got it down and sawed it in half – he was desperate now. It later transpired that Jimmy Walsh, currently our Resident Technical Officer from the Ministry of Aviation, had made it in his apprenticeship with de Havilland in 1926, and it was reported then to have a glue-life of only two years!

At a local Rover Distributor, Doug turned the ends of this propeller to match the variable pitch mechanism on the broken propeller and fitted it. Meanwhile, John Currie, the plane's designer and maker, patched up the wings where they were damaged by Doug's exploits. Within 24 hours, the 50-hour endurance was on again and successfully completed. So, on his way to Elmdon at last, Bellamy flew to the Rover Sports Field, being handsomely beaten by Doug on the road. The Wot would do 120 mph in a terminal dive, and cruised at 75mph, though a head-wind must have made it seem awfully slow. Unfortunately, Bellamy must have found the only imperfection in an otherwise flat field and the plane became stuck in a hole. Nobby Clarke gathered the RGT gang to rescue the plane, but one of the big gates into the Company itself was jammed shut, so the Works Engineers were sought but they had gone home. Undeterred, Nobby and gang, plus chains and Land Rover, overcame the gate and got the Wot onto the RGT hard-standing. Next morning, Nobby confessed his 'crime' to the Chief of Works Engineers who, he was told, would *"have Nobby's guts for garters"*. To Nobby's relief, the Chief said he'd been done a favour. The stuck gate had, at last, been dealt with.

Next day at Elmdon, the Press saw the Wot, now called the Jet Wot, and some Americans were so impressed with the story of how it had evolved that they subsequently bought Bellamy's old Spitfire – somewhat irrelevantly. Bellamy returned south again in the Wot and en route he noticed a small leak (fuel or oil?). Landing opposite a road garage near Henley-in-Arden, he borrowed a spanner, fixed the leak and resumed his flight, getting home in daylight – which was just as well as the Wot didn't have any lights or radio. This, sadly, was the end of the story. Doug found that there was no real market for such a little plane in England so the TP/60 (the new designation for 1S/60 with control for aero use) quietly died.

The TP/90 followed naturally and this was subjected to a 150-hour Type Test in the hope of interesting potential customers. Tony Martin remembers doing experiments with a TP/90 and a Hamilton Standard wooden propeller of variable pitch on the back of a Land Rover and driving round the Rover test track. Just like the time that Doug Lllewellyn went round, it rained and the tree branches again sagged and threatened to foul the propeller. Fortunately, Tony stopped testing before history repeated itself.

Bellamy installed a TP/90 in an Auster Autocrat, and Hants & Sussex Aviation converted a Chipmunk, G-ATTS, which appeared at the Farnborough Air Show in 1966. But it was clear that more power was needed for these sorts of plane and Rover were not interested in going any further, so the TP/90 followed the TP/60 into the grave. "Pity!" – say some of us.

The IS/60 engine and propeller on its first test *(BMIHT)*

The IS/60 engine and propeller with variable-pitch control units departs for the test track in the Land Rover *(BMIHT)*

The Currie Wot single-seat aircraft with TP/60 engine installed in November 1960.

The Rover TP/90 turboprop version of IS/90 developing 116 bhp installed in the Chipmunk.

CHAPTER SEVEN

Boats and hovercraft

Chapter Two mentioned the gas turbine's first operation in a boat, Torquil. Subsequently the Admiralty gave us contracts to install some 1S/60s in their new high-speed boats. Built by Vospers, the Brave Class 50-knot vessels were powered by three Bristol Proteus engines, rated at 3,500 hp each, the first of which, Brave Borderer, was commissioned in 1958. RGT supplied two hand-start 1S/60 generator sets, driving through Twyflex centrifugal clutches. Interestingly, the generators were 220-volt DC 40-kw units salvaged from the moth-balled Princess flying-boat project. We drove a small fan to cool our oil and a double-skin exhaust duct, and our job was to charge the Proteus starting batteries. Otherwise we ran independently of the main engines, though we had an incident when one of our units seriously oversped, seizing our turbine in its shroud. Investigation showed that the Proteus had dumped oil right by our intake and caused our overspeed. Yet another occasion when we proved our engine ran too well on lubricating oil. The intake was altered double quick.

The Proteus-powered Brave Borderer class patrol boats designed and built by Vospers.
(RRHT)

A IS/60 generator set as fitted into Danish Navy patrol boats. (BMIHT)

Vosper sold these boats to the German, French, Dutch, Danish and Malaysian navies, all of course with two 1S/60 generators on board. We called them Specifications 43 and the picture shows a Danish set with a large co-axial duct outside the exhaust, which was fed with alternator-cooling air, through the engine-oil cooler and then outside the long exhaust pipe, avoiding a potentially very expensive lagged exhaust. They also built a simpler boat, the Ferocity class, with two Proteus and two 1S/60 generators, and created wide interest round the world.

In 1963, a potential customer, the Swedish Navy, required boats with a different hull design because the Baltic conditions were so severe. Also, they claimed to need 90kVA output (nearly double that of the other Vosper boats), which meant the 1S/60 could not oblige. Not one for giving up, Doug Llewellyn negotiated with the Swedes and discovered the extra power was required for a self-levelling gun platform. We suggested that the continuous rating requirement quoted was superfluous and that, using our engine and generator rotating masses as a flywheel, we could get the 90 kVA for short bursts within the 50 kVA system we already had. Engine trials confirmed that this was indeed the case and we sold another 15 engines here.

A further development came when Vospers decided to convert the boats to a 60 cycles per second AC electrical system, mainly for much improved availability of spares throughout the world and for some weight reduction. A Maudsley alternator was specified for our 1S/60 to drive, and Doug had quite a job to persuade their chief engineer to use a slightly larger cooling fan, to not only cool his alternator but also to blow through our oil cooler – the difficulty lay in his acceptance of the reduction of his usual 30:1 safety factor on fans to 14:1. To get synchronisation between alternators it is essential that one can very finely tune the speeds to avoid shaft breaking shocks, and Doug contrived a simple means of pressurising our fuel system, which microscopically trimmed the speed. Even the Research Dept said, "Well done!"

About sixty Vosper boats were sold worldwide and our 1S/60 APU received many compliments.

Now, a little sidetrack. The Cuban Navy wanted to buy these patrol boats but it was the time of the missile crisis with the USA. After an initial refusal, they were allowed to buy three Proteus boats, with only two engines fitted, plus quite a lot of spare engines – I wonder what they needed them for?

There was one particularly interesting Vosper boat, bought by a very rich Greek, called Nichiarchos, in which the accommodation was decorated by US and German interior designers. It had a gold-plated bronze drinks bar, hydraulically raised from the floor, and a three-inch thick glass table inlaid with gold. There was also a bronze circular staircase into the passenger accommodation – perhaps slightly over the top? Because Nichiarchos wanted to be able to fly the British Red Ensign, the boat was registered in Britain by employing a British captain. Captain in name, but not in practice. In reality, he had to play second fiddle to the Greek captain who had also been employed and who Nichiarchos wanted in charge.

As the boat was being delivered, they ran into a maelstrom in the Mediterranean, which got pretty rough. To ensure they had sufficient fuel for the trip, they needed to move quickly, so the boat took a lot of green (seawater). Surprisingly enough, the water went down the engine intake to flood and stop our APU. Back at Solihull, Doug received an urgent 'drop everything' call at 9pm on a Saturday. The boat was at Antibes, as they had not reached Nice, and Doug had to scrounge money from his friends and fly out there immediately. Reaching Antibes, he found that they had tried to restart our engine, when it was full of salt, and had bent the starter shaft. So he filled and drained our unit twice with distilled water and did a partial strip (the unit not Doug!). The turbine seemed okay and he fitted a new starter, but he found the fuel pump was not pumping fuel and again discovered our proverbial fault with the governor half-ball off its seat. He restored it and, to his intense relief, the APU again operated. All this work had been in the bowels of the boat with the second APU running to keep the VIPs upstairs comfortable. Doug had been working in a temperature of 120°F.

Two other amusing things happened on this occasion. The first followed complaints which Doug heard that the boat would not quite do its claimed 50 knots and was sitting down too low in the water – the cause of this was found to be ten tons of paint intended for keeping up appearances. The second probably brought colour to Doug's cheeks again - Tina Onassis, who was on board, just happened to join him for a dinner out one evening.

Now, to turn back to prime movers. The first foray was in March 1963, through a man called Carl Keikhaefer, an American friend of Martin-Hurst, the new Managing Director after Wilks retired. He had a 21 ft Vega single-chine boat, fitted with a V6 all-iron engine, driving the propeller through an inboard/outboard unit, which incorporated the reverse gear. Keikhaefer ran this on his personally-owned lake, commonly called Lake X, but which is actually St Cloud, in Orlando, Florida. Rover received this boat and removed the Buick V6 engine to test it on a brake. The rather generous US rating system stated this gave 140 hp – we, more accurately, saw 107 hp. Doug took it to Gosport for a week to measure the performance of both engine and boat and then fitted a 2S/150 with a power turbine governor and a Land Rover clutch, adapted to act as a brake, to permit engagement of reverse without over-speeding the power turbine. We made our own centrifugal fan to cool the oil through an MG oil cooler whilst its exhaust ventilated the engine box.

During trials at the Hamble, a message that Keikhaefer should have the boat demonstrated to him in Poole Harbour was brought to Doug by Peter Twiss (at that time Sales Director of Fairey Marine and who personally also held the air speed record of over 1,100 mph). As there was insufficient time to have the boat transported to Poole by road, Doug set off from the Hamble to make the trip by sea. The weather was lovely and gentle but, as he passed the Needles area of the Isle of Wight, fog descended. Crossing Christchurch Bay at a speed of about 20 knots it was as well Doug had laid out the course as a dead reckoning, a habit he always follows, and he managed to recognise a critical buoy in Poole harbour, leading him to the harbour master. Telephoning Searle on arrival, he was informed of a change of plan. The Keikhaefer demonstration was off and the boat was again needed at the Hamble to show Rover Directors. With a weather forecast of imminent force 8 gales, Doug would have now returned the boat by road. However, a friend of Peter Wilks, being an enthusiastic kind of guy, had other ideas and assured him he could return through the northwest passage, keeping near the Christchurch shoreline. Coming out of Poole Harbour, Doug was greeted by 8 ft waves but, heart in mouth, he maintained his 20 knots and made the trip in under two hours. A coastguard, who had been watching, later told the Poole harbourmaster that he had never before seen anyone go so fast under those conditions. Doug does admit to being petrified. The Board did get to see the boat, bringing protests from Martin-Hurst about the noise of the cooling fan. Doug's solution was to fabricate a simple water-cooled pipe under the hull for the oil and adapt a Hillman Imp plastic fan, driven by a vee-belt to ventilate the engine box.

Rover's involvement with Keikhafer came about because Martin-Hurst wanted to sell him a manufacturing licence for gas turbines to fit to his inboard/outboard units. That is also where the 3.5 litre Buick V8 came from which Rover took over, anglicised, Roverised and sold many of them in the Rover 3500. Land Rover Ltd still market this engine.

Meanwhile, back at Rover, we had subjected a 2S/150 to a 1,000-hour test, which we completed, though with some fuel system difficulties. Martin-Hurst thought Kiekhaefer should also do a 1,000-hour test in a boat. To implement this, in September 1963 another 2S/150, giving 120 hp by RGT measurement, was sent to Florida and installed by Doug into a Vega 23, a little longer but slimmer than the first boat, and started the endurance test. A second, similar, boat with a six-cylinder Chevrolet engine was also there on Lake X and the two underwent the six-day per week, 24 hours a day endurance. There were alligators in the lake but they soon learned to keep out of the way of the boats, probably because one had been killed when another boat rammed them. There were other hazards – moccasin snakes, also called cotton-mouths because they foam from the mouth. They swim like eels and one of them came on board when Doug was in the middle of a job. They are poisonous, but Doug, remembering the alligators, chose not to jump out of the boat and managed to jerk the snake out of the boat with a boathook!

During this endurance test, there were visits by Boeing engineers to see a 250 hp Boeing gas turbine installed in yet another boat. Whilst they were around, the Rover engine was disguised so Boeing did not learn what we were up to. Along with flying off to Nassau in the Bahamas to watch the motor racing there, hazardous flights on a Grumman Goose with fuel-feed problems, helping Keikhaefer's secretary get a pistol through customs, Doug had an adventurous existence during this period in Florida. By February 1964, our engine had attained 356 hours before some cracking was noted on the turbine and generally the engine was not as durable as we had found at Solihull on our own endurance. So the test was terminated.

At Solihull, Paul Langley (RGT designer) took the first Vega that Doug had used at Poole, tarted it up, and showed it at the London Boat Show that year. After the Show, the boat was tested in the big water tank at the back of RGT, splashing a lot of people and bits of Rover! Doug wanted to call her Turbinia but someone recollected that name had been used decades before for the first steam turbine launch, so he called her Turbinia II. It was during this Keikhaefer project that Martin-Hurst told Doug that RGT should concentrate only on selling standard engines and other people should deal with the engineering of installations. That, in Doug's and my opinion, was the beginning of the end of RGT and Rover's involvement in gas turbines.

In 1966 Keikhaefer and Martin-Hurst wanted to demonstrate a gas turbine powered speed boat in the annual six-hour race on the River Seine through the middle of Paris. This would provide wonderful high-profile publicity for Rover but to us the event came to be known as the 'inSeine' race. Aptly called, for it was certainly a hairy race with six hours of a lot of 60-70 mph boats racing one another up and down the river, under bridges, dodging heavy barges, baulks of timber and general flotsam. Not surprisingly, many did not complete the course.

When my wife, Diana, was about six months pregnant with our third child, I was sent to Fond du Lac near Chicago, where Keikhaefer had his main Mercury engine manufacturing plant. Keikhaefer, in making, developing and racing Mercury outboards and inboard/outboards for engines up to 1,000 hp, had become a self-made millionaire several times over. About the time of our interest, the Brunswick Organisation had bought Keikhaefer out of Mercury. This seemed to me to have made him pretty boorish during my five weeks with him but this time was interspersed with moments of humour and charm. With the help of an assortment of mechanics and ocean racing drivers, I installed an annular can 2S/150A in a high speed boat, designing as we went such things as air intakes which could ingest lots of green without drowning the engine, fuel piping, wiring, bilge pumps, etc. Keikhaefer came almost daily and told us how he wanted things and I would then persuade him that my way was the right way and then he would agree – until a few days later. We would then repeat the performance until eventually we had a reasonable working system. This was a tiring and often frustrating experience but the excellent American team were amused, tolerant and helpful.

Talking with some of Keikhaefer's executives revealed that they were not as enthusiastic as he (or Martin-Hurst) concerning the prospects for sales of Rover turbines. They had poor experience of Caterpillar turbines and they knew the engines would be about twice the price of a good piston engine.

We got the first boat going – to around 55 mph – but Keikhaefer decided he didn't like it, or our installation and rang Rover to say he would cancel the whole business. He was persuaded otherwise and chose another boat altogether – a Molinari – bringing more work for us all in a similar pattern. I fitted a fuel header tank system, as in the Le Mans race cars, so no air could get into the fuelling of the engine. We also started work on a third boat – a Luizzi with deep chine – which got everyone really rushing as time was moving on towards the Paris race. Both the boats had 80 (US) gallon fuel tanks.

As a contrast to life at work with Kiekhaefer, there was a pleasant and memorable interlude one Sunday evening when I was invited to join a barbeque he was holding on his 240-acre estate. About 25 executives were there and, to show his great sense of humour, he went round with a large pair of scissors cutting off everybody's tie – except for mine as I protested it was a shame to spoil a nice English tie. He just grunted and passed on. To make the barbeque he and a number of us went off on a lorry to chop down a full-grown hickory tree, which he fed into a chipper which was mounted on the lorry. He put lots of green hickory chips on the barbeque, which smelled marvellous and rather impressive – the steaks tasted terrific too. When we left that evening, he let me borrow his Stingray Corvette car – 0-120 mph in 24 seconds – in which I nearly had an incident (but did not tell him).

The author (third from left) at work on the Molinari speed boat as Carl Keikhaefer looks on.

Over the five weeks I was with Mercury he loaned me one car after another for a few days each – that was a better part of the visit.

I was exchanging cables and letters with Rover as time went on well past the original schedule and Diana became increasingly anxious that I was never coming back. Eventually Keikhaefer let me go, after a trial at 5.30am when air temperatures were about what was expected in Paris in November. The Molinari topped 70 knots and both boats looked very impressive. So now they were off to Paris and I resumed my role as pregnant father.

For reasons known only to themselves, the French organisers demanded that we run the boats on different fuels, so the Molinari was on 'petrole lampant' (commercial paraffin) and the Luizzi on AVTAG (aviation fuel). The day of the race came and Rover Directors were swarming all over the place to watch. Due to the choppiness of the water, we experienced some engine cutting out but both boats got going well and fast. Soon, however, came trouble. The Molinari hit flotsam and was stove in; the Luizzi was broached, hitting the side wall. Frustratingly, the race was won by a Molinari – one of many there – at a speed of under 55 mph from 110 hp. Had they survived, our boats could quite easily have won.

Afterwards, we fished both boats out of the river and got them back to Solihull; flushed out the turbines and they ran satisfactorily. Certainly an 'inSeine' race. To satisfy a query on the gross thrust of the boat installation we rigged up one of the craft over a large welder's cooling tank at Rover and ran the engine. We measured the thrust with a large spring balance mounted on an angle iron frame, and flooded the local area with water churned or thrown out of the tank. We also stopped the Rover welders for a few hours again.

The damaged Luizzi speed boat after the race, on its way to Solihull.

After this display by twin-shaft engines, RGT decided to show what could be achieved with a production single shaft unit 1S/60. They took a small cabin cruiser, called it the Rover Argo and installed a 1S/60 with a centrifugal clutch, a V-belt and hydraulically-controlled variable-pitch propeller (from efficient forwards to inefficient reverse). The Argo was taken to Cowes and demonstrated to a number of potential customers, Uffa Fox having been engaged as consultant. Like many other boats, the Argo was moored on 'The Trots' – big posts down the Solent - but, unfortunately on one occasion, when it was released from the mooring and began to drift quite rapidly, the engine was pushed to full throttle and a rope got tangled round the propeller. Somewhat unceremoniously, we had to be craned out of the water and the rope sawn off.

Another RGT endeavour involved a landing craft – an Admiralty barge – where we installed the turbine at the rear with an upright drive, which could be turned in any direction. Spen King suggested that, in the same way that a Seagull outboard could be turned to get a reverse thrust, this barge could leave the engine stationary but turn the propeller drive below it to any direction. Except for an unfortunate incident where the propeller picked up a hawser and caused our splined shaft to twist dramatically, the system worked but, though the Navy was interested, I suspect we sold no engines.

Uffa Fox in the Rover Argo on the Solvent.

A further exploit in the marine field was to do with hovercraft. In 1963, we sold two 2S/150 engines to the British Hovercraft Corporation, which they fitted to a military craft designated SRN3 but, probably due to the relatively sensitive nature of the design and work, we recall nothing of it directly. However, through a Naval friend, I understand that it was used for a series of anti-submarine trials, using variable-depth sonar, mine counter-measures and fishery protection. At the end of the craft's useful life, she was offered for scrap but not accepted, due to lack of value.

So the Navy decided to use her to discover the hovercraft's resistance to mines. Seven mines, each of 1,100 lbs were exploded; the first some distance away from the hovercraft and the others coming increasingly closer until the final one was just outside her skirts. She was, of course, empty of people and hovering by radio control with radar operating. The final explosion completely obscured the craft with water and, when this subsided, she was still hovering with only some splinter damage underneath. Almost in desperation, the Navy offered her to Gosport Council as a 'toy' for children to climb over. The Council was enthusiastic but refused as she was not considered safe enough. But they did manage to arrange for her to be scrapped.

The IS/60 installed in the Argo, whilst the Vikings look on. (BMIHT)

A 2S/150 unit as fitted to the SRN3.

We were also involved with the SRN4 Hovercraft – a large car-carrying craft crossing the English Channel. This was powered by four Proteus engines, with airscrews, driving the vessel at speeds up to 50 knots. We fitted a 1S/90 Auxiliary Power Unit, which would provide electricity to charge the batteries for starting the Proteus engines, and also air-conditioning and power when the vessel was loading or unloading and the main engines were stopped.

When going on holiday to France one year, I used an SRN4 and, being curious of course, nosed around to find our APU. I did find it, with two rather irate engineers being unable to start it on that occasion. I decided discretion was the better part of valour at that point and left them to it. Yes, we did get to France OK.

I have heard of another 'nonsense' – that someone has bought and fitted three 2S/150 engines in a speed boat involved in a power racing boat competition. It sounds super and quite mad. If anyone knows about this, please let me know more.

The SRN3 hovercraft undergoing sea trials. *(BMIHT)*

SRN3 proves her robustness as mines fail to sink her.

CHAPTER EIGHT

Rover-BRM 1963-65

In June 1962, Bill Martin-Hurst, Rover Managing Director, arranged for the T4 gas turbine car to do a lap of honour before the French Le Mans 24-hour sports car motor race. This fitted in nicely with the advertising we were keen to do of the T4 car and it suited Rover as a build-up to the sale of the coming Rover 2000. It was also useful publicity for the French race and a very nice time was had by all of us who were involved. The car was totally silent in comparison with the race cars and got called the ghost car by the French press. The only thing I can remember about the weekend was my first experience of hearing 'piped' music over loudspeakers at the hotel where we stayed and sometimes this included a lovely accordion.

In November 1962, AB Smith called me to his office to offer me the temporary job of co-ordinating a team linking Rover with Rubery Owen, manufacturer of many car components and owners of BRM. This company made and raced Formula 1 racing cars and it was proposed that they would make the chassis of a lightweight two-seater sports car – actually labelled 'Grand Touring' - to conform to Le Mans rules. We would provide the gas turbine engine and the two companies would jointly run the car at Le Mans in 1963. The engine would be the lightweight 2S/150 referred to in Chapter Four, without heat exchangers, as racing conditions favoured the lightweight, though uneconomic, engine. Somewhat in trepidation, I accepted the offer. First action was to meet Peter Berthon of BRM and discuss with him his general outline of the car, then to visit the Le Mans race organisers, the Automobile-Club de l'Ouest, to discuss the race regulations and our entry. We could accept all their normal rules except for those concerning fuel. We wanted to run on kerosene (paraffin) and needed more tankage in the car to permit decent runs between refuelling stops.

A Special Prize was offered – of 25,000 francs (around £2,000) – for any car, powered by a gas turbine, entered and completing a minimum distance of 3,600 kilometres. At this time, the organisers had no way of rating the turbine against the piston engine, and did not wish to upset competitors with some powerful new prime mover, so this car would not be counted as competing in the actual race. The Automobile-Club de l'Ouest were, of course, delighted that Rover and BRM had risen to this bait and welcomed Peter and me to Paris, where for two days we were entertained and spoiled. They had little hesitation in granting our requirements: the use of kerosene and our need for a 48-gallon fuel tankage instead of the 24 gallons regulated for a two-litre car, to which we corresponded on power levels. Porsche, at that time, were producing around 160-170 hp from a two-litre engine but, as we were not in competition, it was thought that they would have no concern – and neither did they. To make it plain that we were not competing, the car was allocated the number '00'.

One of the older officials of the A-C de l'Ouest was named Gregoire and he introduced himself to us as being another entrepreneur. I could vaguely remember that there had been a French car in the 1920s with, believe it or not, a cast aluminium chassis. I would have been scared stiff of metal fatigue failures but he was very proud that he had made and sold a few cars – it could only happen in France.

In Solihull, we built up a small team, particularly by 'borrowing' Peter Candy from our turbine car fitting shop and George Perry from the electrical area. Then came a visit to Bourne, Lincolnshire, the home of BRM, where we met Tony Rudd, who was the Chief Engineer, and we got on well. As a basis he took the BRM 1961 Formula 1 type 57, as driven by Ritchie Ginther and crashed in the Monaco Grand Prix, up-dated for 1962, plus larger diameter front disc brakes to take into account the lack of engine braking with our turbine. The 'official' leaders of this effort in Rover and Rubery-Owen were Noel Penny – my boss and in charge of all turbine research work – and Peter Spear, Technical Director of Rubery. They generally let us get on with the job.

By this time Rover were working hard on the engine and drawings for accessories, gearbox mounting, intakes, etc; Motor Panels Ltd, part of Rubery Owen, were preparing bodywork in thin aluminium and two big fuel tanks in thicker aluminium; and BRM providing a gearbox which included a crown wheel, pinion and differential unit, two short stub shafts and a rear power-unit mounting bracket. There was also a rather rudimentary reverse gear, as required by the regulations but, because we had no clutch, operation would entail stopping the engine, changing gear through sliding teeth, and re-starting the engine – not something to be done in a hurry. Reverse did not have to be held into engagement (as an SAE report claims[1]) because a cunning rocking-latch had been built into the reverse-gear lever. In order to save weight, the reverse idler gear was machined from nylon. We never had to reverse under power, fortunately, but we did demonstrate at scrutineering that it could operate.

We fitted simple filters of wire frame and porous plastic foam at the air-feed entry points on each side of the car. Because our engine consumed an enormous amount of air, the gravel and bits inevitably sucked in would cause damage. The logic was that these new filter units would succeed when normal filters, in the very limited space we had available, would quickly foul up and block. They were to be changed – which took only seconds to do - at each refuelling stop. We afterwards found that the 'interference' of these filters cost us about 10 hp, so it is just as well that we did not know this at the time. We also placed a small plastic bottle at the rear of the power unit, fed by two pipes, one from the top of the engine oil sump and the other from the separately-lubricated gearbox, to prevent any blown oil reaching the track. It was also a check on what was happening if all was not well within the

---

[1] Spear, P and Penny, N *The development of the Rover-BRM.* SAE Preprint 795B 1964

two areas.

Because the fuel tanks were only about 15 inches deep but over six feet long, we had to develop a fool-proof collection system which avoided any air passing into the engine fuel supply. If this occurred it would cause flame blow-out and a rather long time for the engine to stop and be re-started. A boost pump, of the centrifugal type being used on Jaguar E-types, was fitted at the rear of each fuel tank and this fed into a small supplementary fuel tank of about 1½-gallon capacity through a standard SU float chamber. Either pump could be isolated by the driver turning a switch on the dashboard. With both pumps working, a pressure of around 4 psi was noted on a gauge in the cockpit and, with one switched off, this dropped to 1-2 psi, thus showing when both pumps were working. This was essential as it was found that, even with a large cross-flow pipe between the tanks, one tank would not deliver all its fuel reliably if its boost pump stopped. Also, as each pump started trying to pump air as the fuel tanks emptied, the SU float chamber separated this out and, after this refinement, we never encountered any further fuel feed problems. This system also ensured that every drop of fuel available could be fed to the engine and the 1½-gallon header tank gave the car a full lap with nil pressure indicated from either tank. This may sound complicated but it was really simple, cheap and totally reliable.

I think we were also the first to fix the windscreen water-pipe onto the wiper blade which, the windscreen being well angled, we couldn't 'hit' any other way. However, the drivers still complained that, at full speed (150 mph), the single wiper blade lifted off the screen.

To reduce the unknown where possible, we decided to subject some engines to an artificial 24-hour race on the test bed. We set up windscreen wiper motors, gearing and cams on a test bed so that the engine, fitted with the correct battery, alternator and oil cooler, endured a 24-hour race with as many conditions set as expected – heated intake for a hot day, accelerations and decelerations, pit stops, changed air filters, etc etc. Not all the tests were successful so the endurance was very valuable and, no doubt, contributed to the success at the actual event.

The car went into the wind tunnel at MIRA, the Motor Industry's test track, at night, to allow us full electric power to drive the fans – giving around 90mph. Frank Varker, of Rover Research, was our aerodynamicist and did an excellent job. With a small undertray and a slightly raised lip on the boot, we had no lift at speed. When preparing the car, we knew it was essential that we were ready to attend a practice session at Le Mans on 6-7 April 1963, as this would be the car's – and our – first run 'in anger'. We did run it, briefly, at MIRA in April, having taken the BRM drivers, Graham Hill and Ritchie Ginther, round in T3 and T4 first to explain the characteristics of a gas turbine. Graham and Ritchie were the official BRM drivers that year, though the former was certainly underwhelmed at the prospect of driving the gas turbine car at Le Mans – he would have much preferred a 350hp Ferrari. Due to the banking curvature and, hence, suspension loadings, we could not safely exceed 110 mph on the MIRA circuit; this was quite exciting enough for me as a passenger when Graham Hill got in and belted it flat out around the circuit. Yes, I was scared, and the car bottomed several times, with jacking points scraping the ground and tyres rubbing the bodywork – but we survived. Here is Peter Candy's record of the next few days:

31 March	Engine first run in car at Motor Panels Ltd.
1 April	Car run on Solihull test track. Engine ran roughly as the diffuser bolts had sheared. Engine changed and we ran on the track late that night.
2 April	Car to MIRA driven by Peter Candy, Mark Barnard, Hill and Ginther
3 April	Car left Solihull late in the day for France.
5 April	Team established in very old, tatty, dirty, small garage in Le Mans town. 'Pee Corner' was literally a small urinal in the corner of the workshop!
6 April	Practice in rain. Water came in everywhere – Ginther complained – temporary solutions resorted to. Maximum speed 147mph. Rear end lifting so aluminium sheet attached to vertical rear bodywork, bridging wheel arches. Hill not available.
7 April	Ginther drove first. Hill drove 1 hour averaged 107 mph lap speed.
Post practice:	Modifications to bodywork for wheel clearance; faired in rear spoiler; new windscreen now at 60° to vertical as opposed to 45°; lightening of body support bulkheads and other items, eg handbrake lever, pedals, wind tunnel verification suspension mods.

I recall one difficulty we experienced during Graham Hill's drive. He came in complaining that the engine had cut out so he had to wait till it stopped, re-start it and drive back to the pits. He was not amused; grunted a few lively comments about fixing the problem and stalked off. Quite understandably, he was merciless to the car and I am sure that is why he had the demeanour to become world champion in motor racing's Grand Prix – which, of course, he did. We discovered that the boost pumps were not able to avoid pumping in air with low fuel tank levels and so we 'codged' the problem at the time by filling both tanks full.

The only other incident I recall of this practice was the return journey in my P4 Rover 100. Five of us were on board and, mostly, we cruised at 90+ mph along pretty awful roads. Suddenly, I felt something odd and slammed on the brakes. It was just as well, as Tom Scrimshire in the back, who was a Motor Panel's panel beater was leaning on the handle of the 'suicide' rear door (this is when the door opens out from the forward edge with hinges at the back, the opposite of which is now mandatory) and one of the bumps had caused the door to fly open. We screamed to a halt with two other passengers hanging on to him – half out of the car. What a relief. But then, of course, the door would not shut by over a foot, being bent at the rear hinges. "Stand aside, that's up to me!" said Maurice Britton of Motor Panels. He applied knee, plus his considerable brute force, several times to the bent door and shut it – Click, click! It was amazing to watch - but we tied a rope across the two back doors for the rest of the journey – to make sure. Even Tom laughed about it eventually.

The vehicle engine bed leased for testing 2S/150.

Back at Rover, there were a number of items needing attention in the car and the fuel supply system evolved into the header tank plus SU float chamber already described. An aluminium seat, which dropped into the car driving seat, was made for Ritchie Ginther to reach the pedals – it came in and out as he did.

With this practice over and the modifications carried out, we needed a further run at high speed before the Le Mans race, so we visited Elvington in Yorkshire, an ex-RAF aerodrome with a two-mile long straight. Fortunately, (as far as I was concerned) Ritchie Ginther came to drive. He was generally as fast as Graham Hill, but much more of a test driver, and would describe helpfully what was necessary to improve the car. He went off on the long straight and returned immediately after one high-speed run – much to our consternation. He stopped the car and indicated for me to come into the passenger seat and, unseen by me, winked at the mechanics as I got in. Off we went and at around 130 mph there was suddenly an almighty surge of hot gas down our backs. Seeing I had appreciated his concern, he laughed his head off and drove back to the mechanics. The windscreen shape, now sloped to 60° but very tightly stipulated on height, width, etc by the regulations, had caused a complete reversal of the back-facing exhaust into our cockpit. A gas turbine engine near full power produces a lot of gas and a lot of heat – over 600°C (about 1100°F). We did what we could to remedy this by fabricating an exhaust extension towards the rear of the car – not a pretty sight but effective – and tests continued. On return to Rover, various 'bigwigs' came to view the car and the extended exhaust-stub but the only positive action we could take to 'pretty it up' was to stroke the stainless steel stub with sympathetic brush marks (from Dave Bache – Rover Chief Stylist) I fear we all fell about laughing over that one.

Mark Barnard in the Rover-BRM car at the Rover test track in 1963.

Then came the Le Mans Race proper, when the team went down by car, accompanied by a rather old Land Rover plus trailer to carry the race car. We went to a town called Le Lude, about 10 miles from Le Mans, and into a delightful small hotel which we sort of took over – with much encouragement from Le Maître and Madame.

Now, over again to Pete Candy:

> 13 June   Thursday Scrutineering – traditionally held in the town square. We arrive to find we are the only team there – a change had been made and scrutineering was now at the track, about 10 miles away. Arriving at the circuit in a great downpour of rain, we find the BRM mobile workshop stuck in a flooded tunnel beneath the track. Undeterred, we press on.
>
>   The car is clearing regulations well until we come to ground clearance, where the car has to pass over a wooden block of defined dimensions. It fails the test. The French soft wood block is soaked with rain and has become swollen! Pete brings out our varnished hardwood block made in Rover's model shop and the car passes over – point proved! Good spirited laughs all round.

Practice didn't prove too frightening and both Graham and Ritchie had a good drive. Graham Hill, being of Grand Prix champion material, didn't think much of this tiddly little car run by a bunch of amateurs. Fortunately, we had some super BRM mechanics whom he trusted and we found very helpful and constructive. We were all learning fast about race-car preparation – and each other.

A somewhat facetious example of this preparation was a large sheet of aluminium being placed to cover the sand and drain grating in the pit area, to reduce the foul smell of the drains. Well, you wouldn't have liked it either.

The raised rectangular exhaust outlet on the Rover-BRM.

Our 1963 team was as follows:

Team Leader	Wilkie Wilkinson	(BRM)
Project Engineer	Mark Barnard	(Rover)
Timekeeper	Dr J Jameson	(BRM)
Mechanics	Peter Candy	(Rover)
	Arthur Hill	(BRM)
Refueller	J Sisney	(BRM)

And in the pits we had, as back-up:

Fred Court	(Rover – Engine)
George Perry	(Rover – Electrics)
Arthur Ambrose	(BRM – Gearbox)
Tom Scrimshire	(Motor Panels)

The team in cartoon form as appeared in La Nouvelle Republique, 15 June 1963

    Each team was allowed two mechanics and one refueller, though now in Bernie Ecclestone days, this may be fourteen. Pit passes were numbered metal badges worn on chain bracelets with official 'plombeur' lead seals – happy days! Scrutineering confirmed vehicle weight as 1,725 lb with fuel (1,362 lb dry) and distribution 46/54 front to rear.

 Including the fuel header tank already described, we could hold 48 gallons of fuel and reckoned we should aim for 42 gallons used between pit stops. We expected an engine consumption of about 17 gallons per hour so planned for nine pit stops of around three minutes each, and 2 hour 30 minute runs.

Below, are notes from my 'little book' at each pit stop. Not comprehensive, but interesting. Remember that two mesh air filters were changed at every pit stop, windscreen cleaned, extra seat in and out for Ritchie, and the drivers had to give us warning on data of concern.

Details of the progress of the race, the drivers' reported readings and the pit stop data are given in the tables on page 99.

The only issue which concerned us was noted on pit stop 6 – that the gearbox was hotter to the touch and its oil level increasing. Presumably, engine oil was leaking into the box and, to remove any risk of gearbox failure, we drained it on pit stop 7 and refilled it with correct lubricant. No other problem was noted.

We had done it – we had finished and we felt sure that the car and engine were in a fit state to go for another 24 hours. For the whole race:

Average speed	107.8 mph
Fastest lap (Richie Ginther)	113.6 mph
Total fuel used	372.5 gals
Total mileage	2,588 miles
Fuel consumption	6.9 mpg
Time in pits	33 mins

The power turbine reached between 38,000 and 40,000 rpm along the straight, a road speed of around 140/150 mph, while 680° jet pipe temperature equated to 1,200°K turbine inlet temperature (930°C), which was our engine design point. Graham and Ritchie sought us out after the event and actually smiled and congratulated us, so perhaps the amateurs hadn't done so badly after all! After being up and excited for two days and a night, I recall we were glad to celebrate with a good night's rest before returning to England on the Monday. We were pretty pleased with ourselves, as you can imagine, unlike Aston Martin who had experienced a disappointing race. When we happened to meet in Rouen for a meal on the way home, they very graciously sent over a bottle of Chateau Yquem (the best of Sauternes).

As soon as we arrived home, we had to prepare the car to appear the next weekend at Silverstone. It was the weekend of the British Grand Prix and all our turbine cars were to be demonstrated in honour of the Le Mans result. We got JET 1, T3 and T4 spruced up and, along with the Rover BRM, present at the circuit. To increase the advertising impact, the cars were run in a mock race – with sensible time handicaps so that we ended the three-lap race all together. All went fine except that, with half an hour to go, the Rover BRM would not start. A bit of rapid investigation and we had the fuel pump off and stripped; the governor cleaned and rebuilt, just in time for the track – another close shave. Yes, it was our favourite half-ball governor again!

Spen drove JET1, Ritchie Ginther T3, Tony Worster T4 and Graham Hill '00' (the 1963 Le Mans car). We did three laps and finished together, as planned. Tony recalls that his engine's temperature was definitely too high and he was relieved to finish. Spen admitted that he had tried so hard that he had broadsided JET1 once.

Two interesting events followed soon after: a full performance test at MIRA and a road test by Motor magazine for publication. You may recall that our tyres had rubbed the bodywork the first time at MIRA so, after Le Mans practice, we enlarged the wings a little and now '00' was taken round the MIRA banked circuit at full output, hitting the banking at 147 mph, exiting at 130 mph, averaging a lap time of 1 min 17 seconds. By coincidence, Norman Dewis, the chief tester for Jaguar, was taking his lightweight 'E' type round the same day and doing similar lap times.

T3, JET1, T4 and the Rover-BRM (mock) - racing at Silverstone in 1963.

Later that year, it was decided that we should take a car back to Le Mans in 1964 with the new ceramic heat exchanger engine. BRM uprated the chassis and rear suspension with twin radius arms to accommodate the bulkier and heavier engine.

The lessons learnt from '00' were fed into the design of the 1964 closed-cockpit car and no exhaust troubles or rear end lift problems were experienced. William Towns was stylist and the excellent results he obtained, with exceptionally low drag, can be admired to this day. Also, Noel Penny and Peter Spear had concocted a formula, which became generally accepted, to rate gas turbine engines against reciprocating units. The inlet area to the first turbine was related to the usual swept capacity in such a way that our 2S/150 would equate with a two-litre engine. As 150 hp was generally obtained from two litres, with Porsche achieving about 180 hp, it gave us a real opportunity to compete. Over the years, we got the simple 2S/150 up to 180 hp but the heat exchanger engines struggled to maintain 150hp. Once again, Motor Panels built the bodywork and relations were generally excellent. However, there was one 'nonsense' although we can't recall the reason.

SATURDAY 15 JUNE ----------------------->>>>>>>>>>>>>

Time	pm							am				
	4	5	6	7	8	9	10	11	12	1	2	3	4
Weather	Grey		Sunny		Clear				Cloudy			
AIT °C	28	30	29	26	20	18	18	17	15	14	14	12	11
Position	38	24	23	22	21	19	19	18	18	14	12	10	9
Driver			Hill		Ginther		Hill		Ginther			Hill	
Pit stop			1		2		3		4				
Time in			6.21		8.39		11.17		1.54				
Duration Min/sec			2/50		2/41		3/5		2/46				
JPT °C			680		670		660		650				
Oil psi / °C			25 / 90		25 / 90		27 / 77		30 / 70				
Boost psi			2-3		4-4		2-2		OK				
Fuel added (gallons)			33.7		37.0		40.5		42.0				
Oil added (cc) engine / gearbox			300 / nil		300 / nil		nil / nil		300 / nil				

>>>>>>>>>>>>>-----------------------**SUNDAY 16 JUNE**

Time	am						pm				
	5	6	7	8	9	10	11	12	1	2	3	4
Weather	Mist patches						Cloudy		Rain			
AIT °C	10	9.5	14	15	16.5	18.5	20.5	21.5	23	23	22.5	22
Position	10	10	10	9	8	8	8	8	8	8	8	8
Driver	Ginther		Hill		Ginther			Hill		Hill		
Pit stop	5		6		7			8		9		
Time in	4.27		7.01		9.39			12.20		2.55		
Duration Min/sec	2/41		2/22		12/35			2/11		2/33		
JPT °C	640		630		660			660		660		
Oil psi / °C	28 / 73		25 / 80		25 / 80			25 / 90		25 / 80		
Boost psi	OK		3		OK			3.5		OK		
Fuel added (gallons)	38.5		41.6		39.0			39.6		23.0		
Oil added (cc) engine / gearbox	nil / nil		300 / ¾" above mark		300 drained and refilled			300 / nil		300 / nil		

One day Pete Candy arrived at Motor Panels to find a dismantled Rover BRM parked outside the factory gates, with all its bits. Fortunately, nice words and, probably, apologies restored peace.

Spen King bringing up the rear in JET1.

The 1964 Rover-BRM on test at Silverstone.

The heat exchanger engine was not ready for practice in April 1964, so instead we used one without a heat exchanger, weighted up to compensate. One novelty we assessed this time was the use of variable nozzle vanes in the power turbine. By keeping the fuel throttle on full all the time (on hand throttle), the power to the road wheels was varied by pivoting these vanes. This gave instantaneous torque response at the cost of slightly higher fuel consumption. There was some engine braking effect and the drivers liked the response. However, the negatives of an extra pit stop and, at this stage, the uncertainty of the power turbine blade integrity with reverse thrust, more than offset the very small reduction in lap times. So we returned to a standard vane nozzle.

The 1964 car was improved in detail. As we had gained more space, we used pitot tube air inlets to feed ram air through Purolator paper filters, replacing our previous foam ones. We also introduced a solid-state Lucas continuous sparking ignition unit, called Opus. The large heat content within the heat exchanger made gas generator deceleration very protracted and, hence, the power output decay unpleasantly and dangerously slow. So we stopped fuel supply to the engine sprayer on each lifting of the foot on the accelerator and allowed combustion to cease. Then there had to be immediate relight as the accelerator was depressed again, or the gas generator idling governor cut in, hence the need for Opus.

Peter Candy reminds me that we were always trying to get the best throttle response from the engine, which inevitably meant we were almost into compressor surge on gas generator accelerations. To assist handling, he fitted a tap into the hydraulic feed between the throttle and the fuel pump, by which he could restrict the rate of acceleration to that which did not cause either a surge or 'puke', as he puts it. Puking was incipient surge. Now, not only could an engine be set individually but it could also be reset on any occasion by adjusting this tap, placed near your knee in the cockpit. It was also interesting that hitting a banking, like that at MIRA, could induce the odd pop of surge. Explain that!

Getting the Rover BRM to Le Mans in 1963 had meant that we struggled with a large, heavy and high articulated trailer, drawn by a Series 1 Land Rover, which could attain 50 mph on a good day. For 1964, we were allowed to get our own trailer but it had also to be capable of carrying a Rover three-litre saloon car. This made it longer than we needed and also involved having longer-hinged ramps to cater for our approach angle. As Pete Candy puts it, "Having worked many seven-day weeks and long hours, I protectively drove the outfit to Le Mans for Practice in April 1964 and, when we left to return, I flaked out in the back of the Land Rover leaving another driver to bring us home".

Some time later, he woke up as a swaying Land Rover and trailer hit a tree in a typical French avenue in Northern France. A determined effort by all got the unit back home, though the race car was seriously damaged amidships. The car was rebuilt but the mishap coincided with innumerable problems we were encountering with the ceramic heat exchangers so, reluctantly, the 1964 entry was withdrawn.

By 1965 we had a 2S/150R with ceramic heat exchangers. The drivers were to be Graham Hill and Jackie Stewart and, to give Jackie an opportunity to assess his first gas turbine-powered car, a day was spent at Silverstone in March 1965.

Rear view of the 1965 Rover-BRM, with ceramic heat exchangers, driven by Graham Hill and Jackie Stewart.

Though we now had sufficient experience with the Corning Glass ceramic heat exchangers to say we dared to compete publicly at Le Mans, we were totally unable to predict the life of this unit. Two types of failure molested us: the big bang and the surface fretting. In the former, a ceramic disc just 'cried enough' and broke completely, causing the engine to stop immediately. The life duration could be from minutes to several hundreds of hours and there was no way of predicting that. Corning had tried making the discs in one piece, and as bonded segments, with different arrangements of driving pins on the outside diameter, but all to no avail. A problem was that the ceramic has no 'strain' measurement to judge by. Surface fretting occurred as the air seal appeared to grind away some top surface. The engine would run hotter and need to be detuned. This condition occurred quite often but could remain quite stable – most disturbing. On vehicles we would watch out for 'cigarette ash' from the exhaust and hold our breath!

For this test at Silverstone, the power was set at 130 hp and Jackie completed twenty-two laps, the fastest of which was in 1 min 49.7 secs (96 mph). All readings of temperatures, gas and oil were satisfactory. His speeds were noted at various points round the circuit and compared with an earlier test we had done with Graham Hill:

	Graham Hill with 139 hp	Jackie Stewart with 130hp
Woodcote	108 mph	110 mph
Copse	81 mph	80 mph
Becketts	79 mph	80 mph
Hangar Straight	123 mph	120 mph
Stowe	90 mph	90 mph
Club	90 mph	90 mph
Abbey Straight	125 mph	120 mph

As we became more experienced in the racing world, we learned that we had to test every detail of the car and engine again and again.

The 2S/150R heat exchanger assembly before (upper) and after (lower) failure.   (BMIHT)

The 1965 Rover-BRM rear engine mounting and suspension

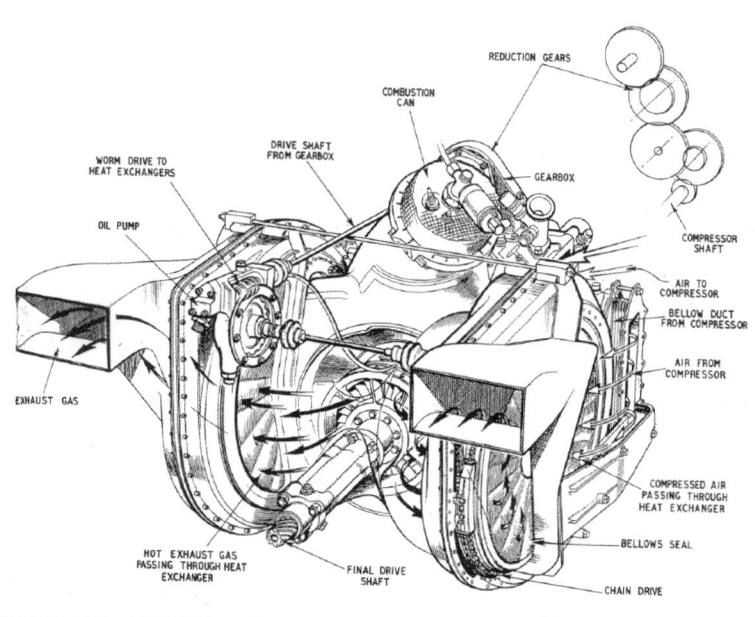

Diagram of the 2S/150R engine   *(Autocar)*

An interesting aside here. When Martin Hurst saw Tony Worster in Spring 1965, he asked Tony if he would like to go to Le Mans to see our second Rover BRM race. Not surprisingly, Tony agreed and was given a Land Rover and three passengers (all turbine people) to go. They had a wonderful time.

On a more serious note, it is now hard to appreciate how the mechanics worked with their backsides in immediate proximity to cars passing at around 150 mph. The team in 1965 were:

Wilkie Wilkinson	(BRM)
John Harbidge and Peter Candy	(Rover engineers)
Dr J Jameson	(Timekeeper)
Arthur Hill	(BRM mechanic)
John Sisney	(Refueller)

and, as back up:

Fred Court	(Rover)
George Perry	(Rover)
Bert Hole	(Rover)
George Dear	(Rover)
Arthur Ambrose	(BRM)
Bill Bentley	(Motor Panels)

One nice little story. At Solihull, Alan Picken, of fuel system staff, was asked if he was going to Le Mans this year and replied, "No, but I am going to Coventry Cathedral to pray!"

Practice seemed to go off all right, though one story Chris Bramley tells shows how all precautions must be taken. As explained earlier, we had adopted a Lucas Opus ignition unit, sparking all the time we ran. However, wearing belts and braces, we had decided to supplement the ignition with our own engine-start make-and-break system. This was mounted on an electric motor with the small vane air-pump to provide air-assistance to the fuel sprayer. Sort of 'just in case', we fitted an extra switch in the cockpit labelled 'Standby' and were later surprised when Graham Hill told us the engine had not been going well till he operated 'Standby' and then all was fine. After that, we gave the Lucas Opus some careful attention.

On Friday before Race day, John Harbidge and Peter Candy ran the engine at base – again at Le Lude – and noticed a slightly increased jet pipe temperature. Due to their concern, Wilkie Wilkinson agreed that they should check the car out on local public roads, this not having been done before.

The test confirmed the higher temperature but that it was stable. One too frequently encountered signs of the deterioration of a heat exchanger, or its seal, was the higher temperature exhaust but, because this was stable, the decision was made to retain the unit for the race, rather than change to the spare engine. Initially in the race, the car ran well, but then the drivers reported that jet pipe temperature was rising further. At several pit stops Peter Candy reduced the gas generator speed to keep jpt below 650° and, at a reduced speed and power level, the car completed the 24-hour race, finishing in 12th place overall and 2nd in the under two-litre class, at 98.8 mph average speed and 13.6 mpg fuel consumption. We were awarded the Motor magazine trophy for the first British car home. After the race, all completing cars were impounded in a 'parc ferme' to be checked for compliance with the race regulations. To our amusement, a Ferrari would not start without having petrol poured into the intakes and catching fire, and a Porsche would only run with a broken-off carburettor held in place by hand, whilst we easily started and ran with no trouble.

Much more interestingly, our engine was now running with an out-of-balance of 200 times the build-level tolerance. We did not know, until we stripped the engine at Rover, that a piece of compressor, including a half vane (alternate vanes between the main vanes), had broken off, probably in the practice sessions and this was the cause of the higher jpt. Another thing we learned later was that our success in completing this race convinced Ivan Swatman, of Ford (USA) to adopt the Corning heat exchangers, instead of his metal regenerators.

When A B Smith congratulated the team, Peter Candy asked if he could drive the car home. Apparently, it was a Jaguar tradition, no doubt much enjoyed by all the team mechanics. However, AB was not to be tempted – though we had no doubts that the car could have done it.

The 1965 car in its racing livery.

Within hours of return to Solihull, Vic Rogers (PA to Martin-Hurst) asked Peter to present the car at the French Grand Prix circuit at Clermont-Ferrand the following weekend. Pete was willing, providing the car and he were flown over as he needed time to change the engine and clean up the car. MH immediately agreed, telling

Peter: "Silver City from Elmdon; 10.00am Friday; Peter Candy and Bert Hole, plus wives; be there" They were!

After the Grand Prix, they motored on down to Marseilles and the Riviera in the Rover BRM, with the transporter following behind. They did various camera shoots on the way and in the mountains behind Marseilles, helping with the publicity for the new Rover 2000 as it was released for sale. In Nice, driving along the Promenade and while shaking off a few pursuers, Peter heard the engine note slowly but remorselessly rising and rising. The tachometer indicated that all was well but Pete turned the engine off to play safe. That was when he noticed a large Air France Caravelle (with Rolls-Royce Avon turbines) taking off right overhead.

A visit to Monte Carlo was next, with a couple of big police motorcyclists leading and going quite fast. The road was somewhat gritty and Peter didn't want that in our engine, so he pulled up close to the motorbikes. They thought this was some kind of challenge and speeded up. Later they pulled off and disappeared. Peter realised that the Monte Carlo signs were indicating that he was now leaving, instead of arriving but he went on travelling west until they reached Menton, confirming that he had missed Monte Carlo. He turned the BRM around and, as they retraced their steps, Peter noticed how all the folk he had seen looking up in the air as he drove past previously, now looked up again. To see the jet?

For officialdom, Peter had to account for the amount of untaxed fuel used during this French touring and he remembers that the Rover BRM did around 10 mpg. He compares this with JET1 at 1-2 mpg; T3 at 13 mpg; and T4 at 10 mpg, for demonstrations.

In November 1965, the car visited the USA to be shown to Corning Glass at Elmira, where it was icy. Pete, locking the brakes, spun the car through 180° - without hitting anything! He had not intentionally done this but the Corning folk were very impressed and applauded strongly.

Visits to various Motor Shows, including New York and San Francisco, meant that the milometer reached 7,000 during this trip.

As referred to earlier, Motor, a popular motoring magazine at the time, decided to review the Rover BRM which was, and still is, a public road car. The road test was reported by two of their journalists: Charles Bulmer, a good friend to everyone who knew him and who later joined us in Rover, and Harold Hastings, who produces good, succinct comments in his articles. This was issued in September 1965 and following are one or two of their observations:

> "Maximum speed around 140 mph and 0-100 mph in 25.4 secs; a best fuel consumption of 31 mpg at 43 mph; and, overall, around 17 mpg on fast cross-country driving; gas generator foot-to-the-floor delay of three secs before full torque available.
>   Bearing in mind the gearbox had no intermediary gears and normal was selected for maximum speed, it was commendable that the car could just start

on a 1 in 4 hill. The brakes were, of course, very powerful but with a lot of pedal pressure, as Graham Hill chose. Road holding was superb but bumpy; steering very responsive; and engine high frequency noise excessive.

To avoid excessive delays in power decay on lifting the throttle, the turbine fuel was shut-off completely and relight occurred (not normally audible) as idle speed was reached. This relight entailed keeping the ignition sparks on all the time."

The author goes shopping after a long day at the office

CHAPTER NINE

Odds and sods

You may think I have a bit of a nerve to use this title for a chapter but it's not so outlandish actually. 'Odds' is, as you will find as you read, a fair title for the little bits included, containing something from nearly all contributors.

The other word in the title is taken as it should be – to mean something from the earth. As Noel Penny and a lot of other people have spelt out, we do need cheaper components to enable turbines to be more attractive financially. Nickel, cobalt, chromium and all the other elements required for making high temperature alloys, need to be replaced with ceramics which, of course, are 'earthy' – from the soil of the earth we live on.

Let us have a go at the Odds. Because they don't easily fit into the general chapter headings, they are identified separately by a bullet point.

* In March 1950, after JET1 had been announced to the public, it became all the rage for vehicular or engine manufacturers to have and exhibit a turbine vehicle. Over the next few years in Europe, Austin, Standard Triumph, Renault, Turbomeca, Fiat and Volvo and, in the USA, Ford, GM and Chrysler, all spent considerable sums of money on gas turbines, most of it for automotive applications. Many dropped out quite quickly, but some hung on with excellent work being done to dig deeper than the obvious twin-shaft concept.

* Chrysler were ambitious enough to make 200 cars with a 130 hp turbine and to make them available to the American public for three months, so Chrysler could gauge the public interest. The rival at that time was the big petrol V8 with automatic transmission, which was very easy to drive and had plenty of power. The net result was that the turbine created generally favourable comments but not so strong as to continue further. Also the Environmental Protection Agency (EPA) emission rules were tightening, so Chrysler decided to concentrate their technical skills on emission reduction work on the petrol engines. This was around 1963. (See Appendix E).

* Ford also had a competitive technical approach and, besides turbine-powered cars, developed the 707 range of big truck engines of 400 hp and obtained experience with operating cross-continent trucks.

- GM delved into the 400 hp turbine, designated the GT250 and GT404. The latter is very similar to the Ford 707 and our 2S/350R with ceramic heat exchangers.

- In Japan, Toyota announced research into turbines in 1987 and soon they, plus Mitsubishi and Nissan, announced that they were working together on a largely ceramic engine.

- Since the demise of the Chrysler turbine car exercise and the closure of Ford's rather public work with turbine cars and big trucks, it has been quieter. Interestingly, the Ford set-up was presented to Teledyne (USA) to carry out US Government-sponsored work to promote advanced materials, pursue ceramic use and aim towards very low emission power units. Pressure ratios rose from our 4.5:1 to around 10:1; turbine inlet temperatures, from our 1,000°C to 1,200°C and combustion systems contrived to produce extremely low emissions, and this helps to keep the concept alive. Ceramics development is a key issue wherever turbine work continues. Interestingly, the downfall of the original Corning Glass glass ceramic - traced to the leaching of the lithium molecules which altered its coefficient of expansion – has been overcome and realistic lives of over 12,000 hours are now claimed for current materials.

- We put a 2S/140 into a Land Rover for cross-country work – the overall unit being reasonably light to enable air transportation. Rather a knife-and-fork job, the engine was in front driving the standard Land Rover axles. Driven over the Rover 'jungle track' (extremely bumpy, over sleepers, through bog, up and down 45° slopes) the unit was superb – very tractable, unstoppable and it even pulled out one petrol Land Rover, which was stuck! The venture came to nothing but might have proceeded if the engine had been available in production.

- Peter Collet (development) was asked to do a study on a larger engine to be used in a tank. He came up with the 3S/200, which was our first, and only, exploit into supercharging our engine. Paul Langley went down to the Fighting Vehicle Research and Development Establishment (FVRDE) to examine the tank environment, space available, etc. and found it all rather compressed and claustrophobic. One excitement came when we tried to run this more-complicated turbine cycle in our test area, using an engine on a test bed, plus a compressor driven by a 500 hp electric motor and big air pipes all over the place. I don't remember it being very successful but it was fun!

* At FVRDE, the Army was interested in fitting a 2S/140 in the Oxford Carrier (a small troop-carrier) to replace a de-rated Cadillac 5.6 litre V8 engine – with similar fuel consumption, and a saving of 380 lb weight. But nothing happened.

* In 1966 the next venture was our Spy in the Sky. We took the 2S/150 gasifier rotating components and high-speed shaft, mounting the roller bearing in rubber rings, with oil damping of course, and fitted a small annular can with three flat sprayers and with air emulsion for start up. To provide thrust, the power turbine section was not used and an adjustable thrust nozzle was placed in the exit from the gasifier radial turbine. This device was intended to power a photographic reconnaissance drone, launched with rocket assistance for a 10-minute flight. The unit, called TJ125, was radio-controlled from a lorry and had to be returned for re-use. We were asked for a life of 25 hours. It was a sharp little unit with a stepped belt-drive to fuel and oil pumps and an adjustable conical centre to tune the engine. As most of the radial turbine's exhaustflow stayed near the outside diameter, this adjustment was not too clever, but it worked. Having built a TJ125 at Rover, we tested it in a mobile frame and it escaped, crashing around outside Project B build shop. Peter Candy took this unit to Montreal, along with the Rover BRM, for demonstration to Century Air Ground Services, who were acting as our agents. Pete fixed up a child's swing, fabricated in Dexion, to demonstrate that the TJ125 would produce enough thrust to power the Drone, called CL89. Noel Penny came out then and they ran the test. Everyone seemed pleased but we never sold them any. Then, after showing it to Keikhaefer to stimulate his interest in selling our units, Noel took the TJ125 to Emperier in Belgium, who also needed it to power a target-drone. They were more excited and bought over 200.

* Always doing interesting things, we designed a single shaft engine, the 1S/100, based on the 2S/75 gas generator, which would have made a very compact unit. Bearing in mind that the 2S/75 had a small annular can with slot-sprayers (ie the fuel was sprayed in a fantail from a sawcut in a round pipe) we inserted a cheaper pipe can, neatly tucked inside the volute scroll. We concentrated on reducing the number of dimensions and increasing their tolerance and this 1S/100 became our 'low-cost' engine, on which Laurence (Tod) Butler and Ces Bedford, of our design team, did much work. Unfortunately, the engine never saw the light of day, though we did succeed in impressing ourselves on the cost reductions possible.

* We once fired-off a cartridge across the air-intake of a particularly sensitive engine, when running, to induce compressor surge, making sure everything hung together all right. It did!

Paul Langley carrying the TJ125 jet unit.

* Another odd one was to see if icing of the intake would hurt our unit. We sprayed in cold water at near freezing conditions and watched as ice built up. When lumps of ice became big enough, bits broke off and, though the engine might surge once or twice, all was well.

* Compressor fouling from oil and atmospheric dirt was more serious and always present. Uncleaned, the compressor performance could drop off 5%, reducing power output. We tried cleaning without stripping the engine by letting it suck in various things like – very small stones;

ground up walnuts; and even a Trichlorethylene spray – all to no effect. The only procedure that we recommended to customers was spraying kerosene and water into the intake as the engine cranked, and then running it. I believe this procedure was also adopted by other industrial turbine engine manufacturers.

* An amusing aside by Doug about the way in which Fred Jones operated. Fred directed the Installation Office and was left-handed. All the drawing boards were fitted with right-handed draughting machines, so Fred looked awkward using them. However, his natural artistic flair came through in beautifully prepared 3D drawings all created from Doug's descriptions – a very useful attribute. Just to make sure that the Project Dept had done their job properly, RGT carefully checked every design dimension and its tolerance to make sure it was as relaxed as possible. I wonder if they found anything?

* In his later years at Rolls-Royce, Fred Morley directed a 'Quick Response' group which could act speedily, without jeopardising any existing design work, to react to difficult questions asked of Rolls-Royce or to create a new engine design proposal. Mechanical design engineer Reg Brealey was the project leader on this team, having started his work in Adrian Lombard's design group and then developed a strong relationship with Morley. He became quite a dab hand on axial-flow compressors and, as he pointed out to me, Rover operated around 4:1 compression ratio but Rolls-Royce are now up to 32 : 1 ratio.

* Amongst their many projects was a proposal that the SRN4 hovercraft could be powered by a Spey engine, mounted in a funnel to avoid shipping any 'green water'; the first APT proposal to employ several Dart engines; and, more recently, considerations of a 56,600 rpm alternator mounted on a 350 hp single-shaft engine. This last would be the power source for a truck with electric motors at the wheels, controlled by a variable frequency converter.

* A particular point Reg made highlights a difference in design approach between Rover and Rolls-Royce. We found we had to reiterate the design process several times when the subject was novel. Rolls-Royce would let the design ideas be shown on drawings rapidly issued on completion. Copies of the final design were issued to the affected areas for comment and action. An interesting difference where both approaches can work well.

Now for the Sods:

Let us tackle the subject of ceramics, which need to include 'cermets' - mixtures with metal. As Noel Penny says, Rover was one of the first to put ceramics into an engine. Besides coating metals to prevent excess oxidation, corrosion or erosion in the 2S/140 variable power turbine nozzles, we also tried various ceramic and cermet materials where they had suitable operating conditions. We used zirconium oxide; metamic (a cermet); silicon nitride (hot-press and sintered); silicon carbide and some glass ceramics, to name but a few. I made a rig, where a wiper motor drove a power turbine vane backwards and forwards between plates of these various materials, in a hot gas stream. Torque and wear were measured.

Ron Hill did a cost exercise where he weighed every bit of a turbine engine, analysed the material constituents and made a clear statement, which appeared in one of Noel Penny's published papers[2], that the turbine materials cost five times as much as those for the same power of petrol engine. That was the beginning of our campaign to find cheaper, high-temperature materials such as ceramics.

To tackle this work, we combined Sid O'Neill, ex Atomic Energy Research Establishment (AERE), Harwell, and Paul Langley to concoct various concepts. An example is silicon powder, mixed with plastic and injection moulded. The plastic is burnt out and then the whole is nitrided, producing a full-size nozzle, which has realistic heavy flanges for mounting. It also has individual blades, with platforms, mounted round the outside at angles to obtain the correct throats. Finally came the shroud, to surround the nozzle unit, which was an isostatic pressing and nitrided. This, in its day, was a remarkable enterprise. Silicon carbide was also tried, having different physical attributes. In 1973, Paul took out a patent on a ceramic combustion chamber.

Probably the most difficult to tackle is the turbine rotor itself. Moving from ductile, fully machined-from-solid rotors (Nimonic 90), through integrally cast rotors with Nimocast 713LC, we tackled twin metal rotors. Tony Martin explored various ways of joining a strong hub material for a turbine rotor (Nimocast 80) with a ring of rim and blades in a very high temperature material (713LC) by electron beam welding. The trouble we encountered with regard to spoiling the beginning of the weld once the circle had been completed, was overcome by puddling the assembly in a frying pan (bought on a Saturday morning exploit) full of molten aluminium – then we got an excellent weld all round. Triumph – but disaster! The welded rim was so thick and heavy now that we could not guarantee that normal overspeed would fail one or two blades, rather than burst the rotor. So we now added axial grooves in the rotor between blades so, effectively, the welding was individual to each blade – and that worked well when the groove was made wide enough. These changes made the welding much easier and we went to much cheaper, low-voltage electron-beam welding. This route can, no doubt, be further pursued.

---

2 The development of the glass ceramic regenerator for the Rover 25/150R engine, SAE 660361, 1966

Ceramic combustion chamber following engine test. (BMIHT)

We also tackled spinning tests on ceramic blocks, representing individual blades and the rim, attached to a metal hub by a root fixing – using a 'de Laval' shape. (This goes back to 1940s in aero engines.) Chris Bramley described this work as "trying to engineer a digestive biscuit". Good point!

So many experiments were made that it is sad to admit that we did not solve all our high cost problems by this form of 'soil mechanics'. The fact that no-one else in the world has succeeded either is only a small solace.

Ceramic components produced at Rover for the 2S/150 engine.

CHAPTER TEN

Leyland and trucks

The end of Rover was coming. The Leyland Group were growing fast – Donald Stokes became Sir Donald and was favoured and helped by the Government to take over the ailing British rump of the car industry. In 1966, British Leyland was formed by the Leyland takeover of Triumph and then Rover. As though they hadn't got enough problems, in 1967 they then added BMC (Austin and Morris, already linked with Jaguar) to form BLMC (British Leyland Motor Corporation). At this stage Bill Lyons of Jaguar said, "Keep off!" and though no master plan was published, we in turbines could see it might be the end of our work. However, noises were made which suggested that a turbine engine for a heavy truck might be considered and Noel Penny made presentations to the Stokes' Board.

Along with Leyland, we prepared a Comet tractor to accept one of our ex-race 2S/150R engines fitted with a four-speed semi-automatic gearbox, though we did notice that the Leyland gears were noisy by our standards. This box, produced by Self-Change-Gears (SCG) used pneumatics for operation, controlled by some rather basic Post Office electrics and relays. Once, doing 60 mph on test at MIRA at full weight of 32 tons, the gearbox decided to change all the way down the ratios, 4-3-2-1, with the result, of course, that our power turbine grossly oversped, grew and friction-welded itself to its shroud. Not dangerous for people because the overspeed was fully contained, but expensive and tragic for the wrecked half of the engine! With the engine rebuilt, demonstrations were made to Leyland in Centurion Way and the gearbox repeated its failure: 4-3-2-1 = seizure! The experts in this technology reckoned that a mechanical resonance – like running over ridged concrete – caused the failure. The electrical gear shift nonsense was converted to a Lucas/CAV Hydro-mechanical controller and the trouble was not repeated.

The next incident happened when the Comet was driven down to London to appear on the Tomorrow's World TV programme. On the journey, the engine temperature was noted to be rising and 'cigarette ash' was seen in the exhaust indicating a partial failure of the ceramic Corning heat exchanger. It was also snowing (the weather!). We decided to press on with the dangerously hot engine. The demonstrations were completed and the Comet managed to limp home to Solihull but it was a close shave.

In August 1967 the BLMC Board decided to retain our Turbine Research Group, now called LGT (Leyland Gas Turbines), to design and put into service a gas turbine engine of enough power for the largest tonnage trucks expected in the future – at that time 32 tons was maximum but 38 tons was anticipated. A power level of 350 hp was targeted, with stretch to cater for the legal imposition of a power/weight ratio of 10 hp/ton, which was anticipated. So the 2S/350R became our whole work; the design based on a scale-up of the 2S/150R.

The presence of RGT at Solihull was unacceptable to the BLMC Board; their building was needed for the truck engine work, so Noel Penny negotiated with John Parkes of Alvis to take over all RGT commitments. This also meant that we were able to gain some RGT folk, who did not want to move and would help make up losses we had suffered when Leyland took over. We retained a good team. At a later stage, BLMC wanted RGT out of the Corporation and Noel negotiated with Lucas Industries to buy RGT in 1972, where they still are today.

Ron Ellis was the BLMC Board Director who controlled our operation and Dr Albert Fogg was their Technical Director, who became my specific controller.

The whole attitude of the BL Board to gas turbines was, as with a new diesel engine, that we just had to design and prove the unit so that customers would be encouraged to buy. The necessity for 'development' of the brand-new engine did not feature in the Board's approach. This was very much to our undoing as life proceeded.

Leyland Comet tractor unit powered by the 2S/150R engine.

Though we naturally designed the 2S/350R as the best unit we were able, in the light of our experience with the smaller units, we had never got the ceramic heat exchanger reliable. The sight of 'cigarette ash' (debris from the heat exchanger surface failure, which was often progressive) or a big bang with complete disc fracture were still very much with us. At Rover we had achieved several 1,000+ hour-lives of heat exchanger discs, but the norm was less. We learnt later that, by 1973 Ford had seen discs attain 4,000 hours though, again, many lasted much shorter periods. The 2S/350R was of twin-shaft design with moveable power turbine nozzle vanes, twin ceramic heat exchangers of 28-inch diameter, cast main air casing with insulation by moulded Foseco blocks. Stokes gave us eighteen months to produce six ready-to-use engines and to demonstrate one in a truck to Anthony Wedgewood-Benn, then Government Minister for Technology.

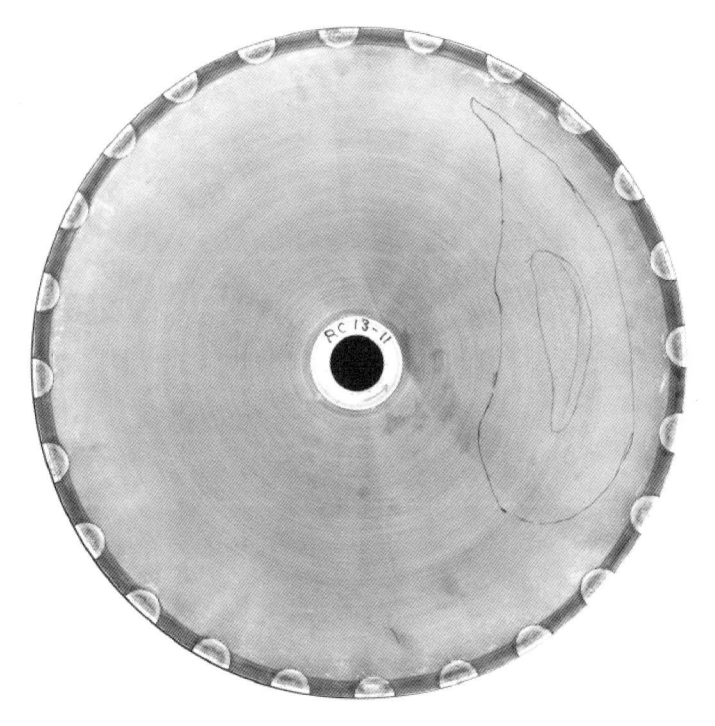

The 28-inch diameter heat exchanger disc with inserts for the 2S/350R engine. (BMIHT)

Ces Bedford made full-scale models of the complex and, for us, huge main casing to be cast in Spheroidal Graphite (SG) iron, a strong and slightly ductile cast iron. He also made models for the engine installation under the tilting cab of the new Marathon tractor to ensure everything was cleared. (We were peeved to find that the new V8 diesel Leyland were going to use needed a lot more space than our engine). These models, coupled with twelve sheets of drawings, helped the manufacturers, West Yorkshire Foundries (Firth Alloys), for whom this was their largest job, and everyone was delighted to find the first casting was successful. To test the large ceramic heat exchangers at realistic temperatures, Chris Bramley built a rig with an LPG (Propane) heater to simulate the exhaust. The first trial was memorable! We tried lighting up the gas system about ten times unsuccessfully before discovering that the on/off valve needed to be set the other way up. When light-up finally occurred, we had the most enormous explosion any of us had ever heard. One observer reckoned that a large bellows we had in the exhaust system moved five feet and there was dust and asbestos everywhere.

The regenerative heat exchanger test rig, instrumented to measure heat transfer, pressure and mechanical losses in a complete disc assembly.

So, in September 1968, we assembled the first 2S/350R – as always, just in time. The unit was fitted to a prototype Marathon truck and we drove it to AEC's place in Southall. We had to drive it up a ramp (outside the Commercial Vehicle Show) to demonstrate it to the Press and, as we climbed the ramp, we saw to our horror some 'cigarette ash' from the exhaust. Only a few other folk noticed and those who did were too tactful to mention it (whilst we had kittens!). The truck got a good report – the high power level target was clearly popular. We built three trucks for ourselves for 'proving' and another three as tankers, for Esso, Castrol and BP's use and for publicity.

In Chapter Nine I referred to the use of silicon nitride to manufacture turbine nozzle vanes and, in these Leyland days, we embarked on a rather ambitious programme to create our own heat exchanger disc to avoid the monopoly held by Corning Glass in USA. We employed three individuals from AERE Harwell who were trained in, and understood, ceramics. Eric Silverstone, Sid O'Neill and David Noble, doctors all, joined us and were quite an education for us mechanical engineers – though maybe we were for them, too!

The laden Marathon truck and installation of the 2S/350R. *(BMIHT)*

They helped us assemble a 28-inch diameter heat exchanger disc of silicon nitride, for which we obtained financial help from the MinTech. The material retained its normal thermal characteristics, namely a low expansion as the temperature rose, but we had experienced such good results in other hot experiments that we, supported by others, dared to hope that it would be strong enough to avoid fractures. It was not, unfortunately. The disc fractured (big bang type) once subjected to full gas temperatures, bringing about the conclusion of that venture.

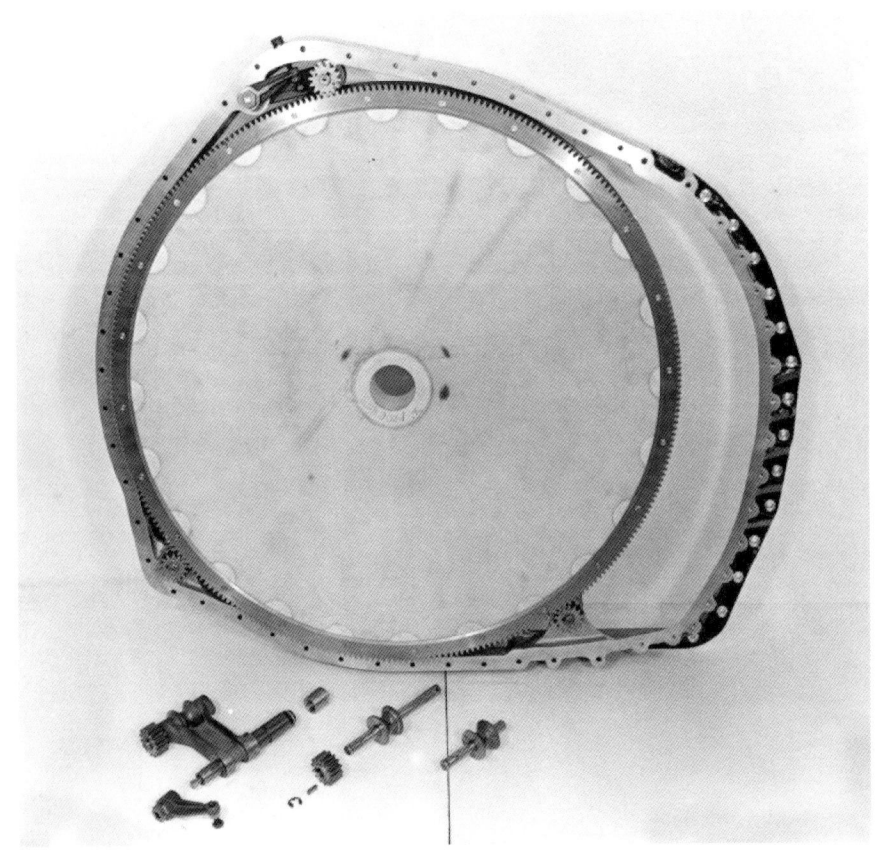

The 2S/350R heat exchanger disc mounted in its cover

In 1968 the BLMC Board asked Rolls-Royce to examine the proposed LGT plan to develop the 2S/350R to meet production commencing in 1972/3. Rolls-Royce visited LGT in November and issued a thorough assessment of the engine design points, the likelihood of achieving these and the programme planned. A number of the engine design points were identified as important and these are given here:

Weight	980 lb
Temperature range	−30°C to +50°C
Altitude range	0-1000 ft
Factory cost (including tooling) at 1000/year	£3.2/HP
Engine life TBO	12000 hrs
Turbine inlet temperature	1067°C developing to1200°C

The 2S/350 engine core

Later that same month Rolls-Royce gave general approval of the engine design but were cautious on the planned time scale, noting insufficient time to rectify problems arising. They also visited Caterpillar Tractor Company in the USA and their report identifies assistance they would need to compete in the same market. Rolls-Royce examined the world-wide market for engines in our power range, agreeing there was excellent market potential. Their observations were quite favourable but they were only too aware of the considerable risks attached to the time scale and, interestingly, not spelling out the biggest technical hazard – failure of the ceramic material for the heat exchanger.

In November 1969 Leyland Gas Turbines, assisted by financial specialists, made another assessment and concluded that, "The major objectives involving large quantity production in 1972/3 are realistic in light of the new expenditure outline".

This reflected a small increase in the engineering budget for which we had pleaded and the satisfactory build-up of engine endurance hours (which we were achieving by working very long hours and replacing failed components promptly). The document is couched in pretty encouraging terms – too much so in my opinion. It also mentioned that MinTech had agreed to invest £1.5M if progress was deemed satisfactory and that appeared secure. Other Government involvement was referred to, namely the National Gas Turbine Establishment (NGTE) and AERE (Harwell), the former for regular turbine consultancy and the latter to support work being done to create our silicon nitride heat exchanger.

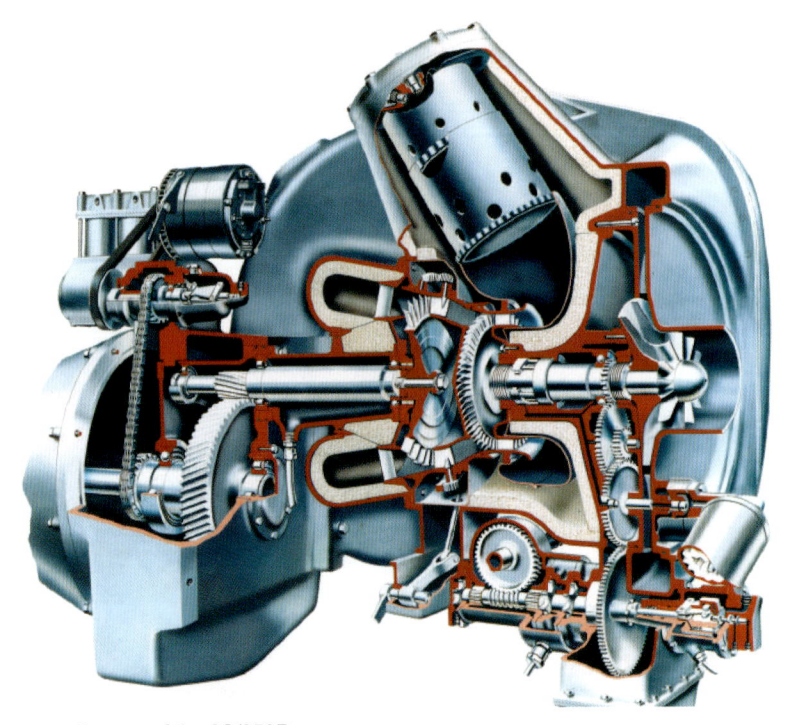

Cutaway diagram of the 2S/350R

A note circulated in Rolls-Royce in August 1970 explained that Ron Ellis had turned down their terms for technical assistance on the 2S/350R and was generally cool on collaboration with diesel engines. Rolls-Royce had been assured that they would be able to purchase silicon nitride heat exchangers if they wished, but BLMC would have to approve at the time. Altogether a frustrating episode, though it is unlikely that Rolls-Royce would have overcome the ceramic problems that quickly.

The gearbox, using the excellent engine torque characteristic, was a five-speed

SCG (Self Change Gears) automatic unit with hydraulic controls. The truck, light or fully laden at 38 tons, was superb to drive – smooth, manoeuvrable, holding speed well on a grade and in pulling away. Though heat exchanger reliability was always a problem, regular truck drivers much appreciated driving them. The engine and gearbox shared the same oil, though we found that their filtration requirements differed. Peter Candy recalls that the fibrous bits of gear-band material were not good for the control valves and caused trouble once or twice. On one occasion, we experienced fifth gear and reverse engaging together, with the prompt cessation of fore or aft motion – it was on the Coventry Road in Sheldon (a very busy road), effectively blocking it for rather a long time! This occurrence also showed something which might have saved an engine one day – the helical power turbine pinion was shrink-fitted on its shaft and the brutal stopping of two gears pulled off the pinion, probably preventing worse damage.

The 2S/350R engine with five-speed gearbox   *(British Commercial Vehicle Museum)*

Another problem in the installation was that our idling fuel flow was too high and part of that was traced to the power absorbed by the inefficient brake air compressor. John Hughes, a consultant we engaged, devised sensible improvements to it and we fed our own compressor delivery to the brake compressor's intake. This combination enabled us to fit a smaller brake compressor, so reducing power loss considerably, and our idling consumption a little.

We used another piece of standard diesel equipment, a powerful 24-volt Butec electric starter. As it was engaged, the considerable shock-load gave us the 'willies' and eventually broke off gear teeth in the gas generator drive train. To overcome this we invented a soft-start technique. With a fixed resistor put in the field (ie the low current circuit) when the starter circuit is engaged, the starter just gently turns and picks up the slack. After half-a-second, the resistor is shorted out so full battery power is unleashed and the gas generator accelerates normally. This system worked well and made us a lot happier!

The engine was remarkably trouble-free, considering the novelty of the design: we had generally covered most of the unknowns in the 150 hp engine and only embarked on a few improvements for the larger unit. The heat exchanger drive started with the original chain drive concept with ceramic pegs round the periphery of the disc but we changed to a simpler system, with a single gear drive, mounted on S-shaped springs round the disc and with carbon bearings for the shaft. The cold-side air seals were fabricated from glued-on blocks of carbon, but the hot-side seal was sprayed-on nickel-oxide as before.

Paul Langley remembers a somewhat outrageous task that we tried – welding 0.010-inch stainless steel fingers to a 0.5 inch-thick seal for the heat exchanger. Our virtuoso, Jack Hedricks, managed it and it was also he who sprayed nickel oxide on high temperature seals – a skilled task!

Spraying nickel oxide onto the 'hot side' of the heat exchanger seal.

I believe that if the ceramic heat exchangers had become reliable Leyland would have pushed on – they were very publicity conscious and, when going well, the turbine trucks were excellent advertising.

The 2S/350R had a Lucas electrically-controlled fuel system which had good and bad points. Good, because we could now adjust the variable nozzle settings to improve part-load efficiency and we chose to go for a constant T6 – that is the gas temperature between the two turbines. This was quite an improvement over our earlier work on the car engines. The not-so-good, and sometimes a real pain, aspect was the solid-state electronics used. These were circuits and components etched on thin-film tiles, set in a box and potted with silicone. This would be fine for a production unit, but every time we needed to alter any characteristic, a tear-up and re-make became necessary. This reflected the Leyland expectation to be able to move rapidly into a production state and it made our development work correspondingly more difficult.

For many years we pursued the goal to cheapen the turbine engine and had formed a Value Analysis (VA) section, which investigated the advanced methods of manufacture that were becoming available. Casting of hot components, like nozzles and turbines, involved work with a number of firms: Deritend Castings, Rolls-Royce, Hepworth and Grandage, and Centrax and for a while Ces Bedford and Ron Hill lived in those firms. We also explored ECM (electro chemical machining) and spark erosion for moving difficult materials. Ces nearly persuaded the firm Schenk to balance our shafts automatically by spark erosion. In truck days, Ces Bedford had to liaise with the VA section at Leyland and we were quite pleasantly surprised to find that the truck world was relatively inexpensive compared with our automobile environment, though the turbine was still too dear.

At about the same time as our truck engine work was progressing, the high technology part of British Rail (BR) was designing a fast gas-turbine-powered train, called the Advanced Passenger Train (APT). Having initially considered a Rolls-Royce Dart driving a complicated arrangement of mechanical drives, through gearing to two large bogies of six wheels each, BR had second thoughts and changed to a single large alternator, with motors in the bogies. Then they had their third change of mind and got us involved as the potential provider of commercially-viable gas turbine engines, in place of the Dart. The positive publicity of our involvement was attractive to Donald Stokes. However, the prospect of our development team providing ten 350 hp engines for this train made us boggle with disbelief. Because the large ceramic heat exchangers were still very much our Achilles heel, we undertook (reluctantly) to provide BR with a batch of engines without heat exchangers. This for us in 1971, of course, meant developing a fuel system, combustion system and quite a lot of detail unique to BR and the manufacture, delivery and servicing of these engines – over and above all our own truck commitments. But somehow we did it! BR had a fourth change of mind to eliminate the complicated mechanical driving system, and each of our turbines drove alternators powering electric motors fitted to all axles of the APT bogies. Of the ten engines we provided, all of which were coupled to alternators, eight drove the power motors in two bogies, and two worked as Auxiliary Power Units, providing air conditioning, battery charging, lights, etc.

The 'raison d'être' for the APT was rapidity on short journeys, and the technology chosen was the extensive use of aluminium to keep down weight and a tilting mechanism to permit 150 mph cruising on standard BR rails, designed for 90 mph. The low weight of the train would, therefore, not load the sleeper and rail mounting any more than they had been designed for. But to help passengers cope with the high cornering forces involved, the whole train would tilt, so people only experienced a very slightly greater down-force (vertically for each person). Certainly I have now experienced the view of the horizon moving up or down through the windows, at 140 mph, and found drinking a cup of coffee quite normal. Unfortunately, this tilting did not always work properly and must have caused distress to a number of test engineers.

A 350 hp engine used on the APT.

Phil Gardiner became our APT engine project engineer and tells two stories, which may (or may not!) be true. The first concerns moving the train from a shed, where work had been carried out, to the track just outside. There was a slight down-slope on the rails going out and the APT was moved outside by running just one of our engines. The process was repeated in order to return the train to the shed. However, the harder work to move up the slope caused wheel-spin and it was found that two of the driving wheels had lost 1/16" x ½" wide metal. Wheelspin is not a good idea! The second of the stories is by hearsay and a bit terrifying. Near to the end of the project, a bump/hump was noticed on the outer skin of the train. On investigation, it was found that one of our engines had burst a power turbine and a piece of metal had embedded itself in the side of the train. I am grateful it did not go farther.

A general observation on this train – it was officially designated APT(E), where E meant Experimental and means 'Proof of Concept'. However, the Government and the Press assumed that this was the Prototype APT, which it certainly was not, and so it received probably more than its fair share of criticism for its failures and problems.

The British Rail APT(E), propelled by ten non-heat exchanger Mk 1 engines through alternators and wheel motors on four axles. The first run was in June 1972. Speeds of 135 mph were achieved.

In 1972, as part of the ongoing attempt to reduce the cost of our engine, and because Leyland wanted a different shape of unit to suit their Marathon chassis, we decided to go to a single-heat-exchanger design. Designated Mk 2, this would still use a 28-inch ceramic disc and suffer the slightly lower thermal ratio and gain improved compressor outlet scroll and exhaust duct shapes.

Our good experience with the last 150 hp engines at 950°C turbine inlet temperature, 1,000°C for 2S/350 units, led us to aim for our highest ever level of 1,050°C for these Mk 2 engines. On one run, we attained a lowest fuel consumption of 0.43 lb/hp hr, which compared with a best diesel engine figure of 0.33 lb/hp hr. Our Marathon attained a figure of 5.8 mpg unladen.

We did have a few 'purists' among us who said we should have kept the twin heat exchangers, from which we had once obtained 0.40 lb/hp hr. Instead we had a cheaper engine of more convenient installation and achieved 400 hp. We also completed an excellent 1,000-hour endurance test, though I had to plead for a few weeks' stay of execution on the closure of our turbine department to allow this to happen. So in 1974 when the BLMC Board decided to stop all gas turbine work, the end, under Rover auspices, really had come. Lucas still carry out service and repair work on RGT units and Noel Penny set up Noel Penny Turbines Ltd in Coventry – and there's another story to be told some day.

We reckoned that, in spite of the heat-exchanger problems, the oil company tankers covered a total of 75,000 km, which is not bad for a 'publicity exercise'. In addition, the APT reached a speed of 152.9 mph and the engines accumulated 5,000 hours. Again, not bad. One more thought – during this 2S/350R development programme, we heard that the around-the-world failures of Corning Glass ceramic heat exchangers had caused the American Automotive Industry to pull out of their direct work on turbines. I have no doubt that knowledge helped Leyland to pull out, too.

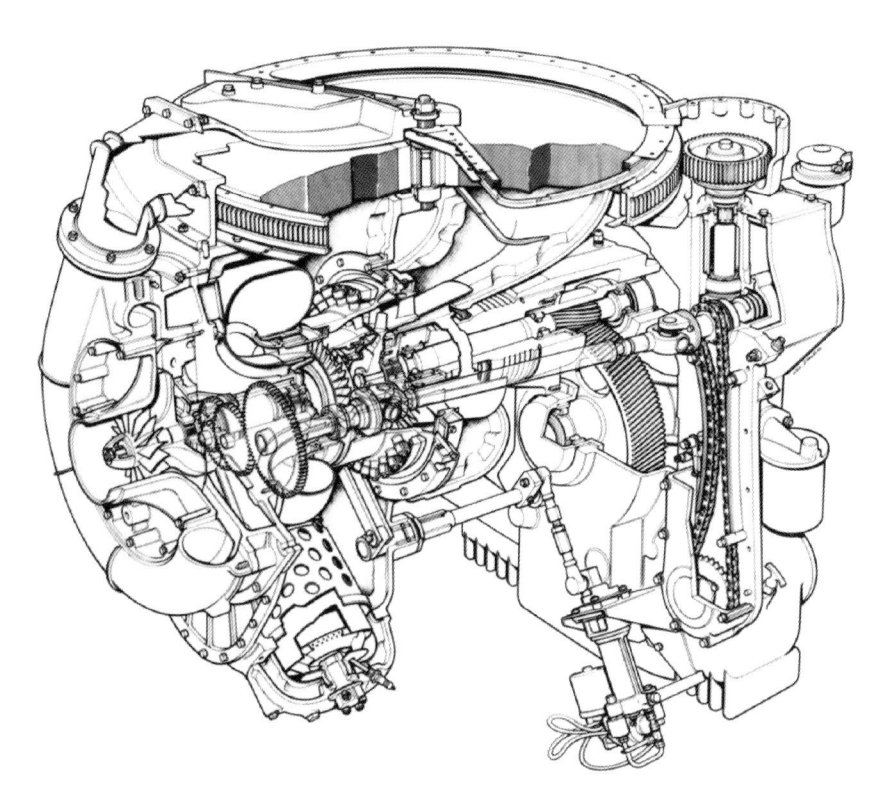

The 2S/350R Mk 2 engine

CHAPTER ELEVEN

Is this the end?

Speed

I've had to slightly tweak the facts to suit the truth …. but the following is quite interesting. You may have noticed, if you have read thus far, that the figure 150 mph figures a few times.

The first is with JET1, which reached 151.9 mph in Belgium in 1952. It may be true that the aerodynamics were rather crude and the engine was screwed up as far as was dared (no doubt) – but we did it. Secondly, in cars, came the Rover BRM whose top speed was noted at 150 mph. Thirdly, the Advanced Passenger Train reached 152.9 mph. Although full of problems and too technically risky for the railway practice at that time, it was a genuine achievement – and, alone, we did it. Next comes a slight concession, referring to an aircraft (which would normally have no difficulty in achieving the great speed of 150 mph). The Curry Wot (biplane) would cruise at 75 mph – half my desired speed. And the final concession is on water. The Molinari we powered for Keikhaefer was measured at 70 knots, and that is near enough 75 mph.

People

People are useful, and even necessary sometimes! Our gas turbines – clever though they were (are?) – still needed people at various times to make their bit of history.

To go back a bit - to the 1940s - when Ces Bedford remembers Helen Street in Coventry before the Blitz; then his move to Chesford Grange, near Kenilworth, and then to Lode Lane, Solihull. He reckons that, under instruction from Geoff Seale, (Chief Designer – Rover), he made the first arrangement drawing for the 'new vehicle' – the Land Rover. (Nothing to do with turbines, of course, but interesting).

Spen King was the boss and Noel Penny joined in 1950 and found, as he puts it, a peculiar grouping of people – draughtsmen, technical assistants and fitter-testers. Co-ordination was superb and most people seemed to be able to turn their hand to be 'jack of all trades'. Noel recalls the 'Black Hole of Calcutta'; a name given to three small offices between the drawing office and the technical office.

I was the second graduate accepted by Rover and was sent down to Project Department (Gas Turbines). The first by a few months, Bill Kennard, tells this story:

> "When I arrived on the scene and joined John Garrett, he was only too delighted to find that I was familiar with the repetitive calculations based on the annular ring concept. Here, the disc was sliced (in theory) into a large number of thin, annular rings where the boundary conditions at the hub and the blade roots were known. I was handed a cylindrical slide-rule, equivalent to an 18-foot-long rule and told to get cracking and calculate! My memory is a bit vague on the time it took to do these calculations, but I seem to remember that it was of the order of three months and then along would come Spen on his crutches (skiing accident?) and, using a crutch as a pointer, would either spot an error or decide on a new profile for the disc. A black smudge from the rubber tip of the crutch reminded you just where and how to go about the necessary mods. Just as I was leaving to go to Birmingham University, I think that Rover invested in a computer, which would love that sort of calculation so I do not suppose that anyone had to suffer that agony again. It was all good fun and, for an engineering graduate, there were challenges aplenty - to design a bearing support that would flex but whose natural frequency of oscillation would be outside the working speed-range; etc."

I developed a sort of warning system when Spen visited our office and might be obscured by the drawing boards. I would loudly proclaim that "a dimension was X, where X always included 1/8 inches" and the other folk knew Spen was in the vicinity. It worked a few times.

A snippet from Nobby Clarke:

> "One lunchtime, Harold Smith and I were working flat out to fit the new hand-start mechanism for a Neptune (before it became 1S/60) in time to demonstrate to Sir Harry Ricardo, who was visiting. We just got it done in time and tried a start with Fred Hulse helping. We tried three times but it would not go and the Fred looked up inside the intake, smiled broadly and said that the engine was rotating the wrong way as we wound – and in came Sir Harry. Such timing. And he thought it hilarious, too. He then admitted that his firm had once designed a client's new engine with the camshaft turning the wrong way!"

In 1957, Spen forsook us to pursue Rover New Car Design and Noel took over as Chief Project Engineer. He also was appointed Technical Director of Rover Gas Turbines Ltd in 1959 and Managing Director of Leyland Gas Turbines in 1966. Sir George Farmer, for some time Chairman of Rover, also became Chairman of RGT; A B Smith taking over the role later.

Down in turbines again, very important people (like me) arrived in 1953 and I now tell a story against myself, which certainly 'put me in my place!'

Spen, Ces Bedford and I were having an intense discussion around Ces' drawing board when Ces suddenly said, "Nobody, but nobody puts their foot on my stool!" Spen said, *"That's told you!"* I removed the offending foot and didn't know where to put myself.

On another occasion, when I was wearing a smart blue suit I had just bought, along with a blue shirt and tie, Spen strode into the Technical Office, saw me and said *"Ah! My little boy in blue!"* and then walked out. I was furious but too intimidated to retaliate.

Doug Llewellyn made an interesting observation: *"The people involved in our gas turbine work were clearly not there for the money. We all knew that Rover did not always have the best-paid jobs. But the mixture of apprenticeships and degrees, plus innovative management, seemed to encourage those who had a flair for turbines to flourish and, altogether, we did a lot of new things and enjoyed the work."*

I go along with that, too.

When we experienced the Leyland takeover, Ron Ellis (see Chapter 10) was keen to move all our Research Department up to Leyland (he wanted our drive and expertise on his doorstep.) In spite of a whole day of apparently unsuccessful discussions with Sir Donald Stokes to prevent it happening, in May 1971 Noel resigned his post – and we stayed in Solihull.

Noel told me later that Sir George Farmer had sold licences for our fuel system, bringing in £75,000 and we always found that any patents we in turbines sought were encouraged and obtained.

At the height of our work on the 2S/150R in 1966, we had two important individual visits. One was from Fred Morley, Director of Design of Rolls-Royce, Derby, and one from Lionel Haworth, with a similar post at Rolls-Royce, Bristol. Noel and I spent all day with them and they both made similar statements. These were to the effect that, before they came they knew they would do a lot of things differently – particularly on the heat exchanger. By the end of the day (8pm on both occasions) they had changed their minds and agreed with what we had done, endorsing all our technical approaches. That certainly made us feel better.

When Rover and Triumph were amalgamated with BMC, the Austin gas turbine department was closed down. They had made a car engine, though it was a bit ponderous by our standards, and they had made a number of commercial turbines to sell. Dr John Weaving had been in charge and now joined my staff, bringing plenty of wisdom. Peter Parker was, and continued to be, a brilliant aerodynamicist and developed superb compressor designs, better than we had at Rover. The 2S/350R particularly profited from his skills, with good compressor efficiency and surge-free transient behaviour. A very bright development engineer we had, Alex Ritchie, turned his hand many ways at Rover and he was a real tonic for us. One special story was of the occasion when Alex borrowed Joe Poole's brand new hire car for a business trip. He commented to Joe on his return that "the bonnet flapped at speeds above 85 mph". Joe understandably went berserk. Always one for doing outrageous things, Alex became quite famous when ballooning with Richard Branson. On a 'round-the-world' attempt, Alex climbed onto the roof of their capsule in darkness, to jettison heavy fuel tanks and, Branson said, "saved our lives!" Soon after that, he was engaged in a skydiving exercise when his parachute failed to open and he fell, breaking nearly every bone in his body. Sadly, he died soon after and his son commemorated Alex by calling him a real fighter.

For your interest, and generally on public view if you feel so inclined, are many of our vehicles and engines.

Aero engine

Rover-built B26 ST4, together with a Power Jets W2/700, is at the Midland Air Museum, Coventry Airport.

Cars

JET1	at the Science Museum in London and still looks good. Non-runner.
T3 Base Unit	at the Rover Heritage Museum, Gaydon, and fitted with an Aurora engine. Non-runner.
T3	at the Heritage Museum, Gaydon, and also fitted with a non-heat exchanger Aurora engine. Non-runner.
T4	at the Heritage Museum, Gaydon, and fitted with a 2S/140 engine, which is too hot to run, probably due to excessive high pressure air leakage from the heat exchanger.
Rover-BRM (1965)	at the Heritage Museum, Gaydon, with a working 2S/150R engine. Having recently been repainted, the car looks really good.

Trucks

Comet with 2S/150	at Leyland in storage. In tatty condition and the Museum hopes to restore it soon.
Prototype Marathon	also in storage at Leyland, but its engine is on display at the British Commercial Vehicle Museum, King Street, Leyland, Preston.
T37 Marathon With 2S/350R Mk 2	is at the Museum of British Road Transport, Hales Street, Coventry.

Rail

APT is at the BR Museum, York. Not on display, but can be seen by arrangement.

APT engine is on display at the above Museum.
2S/350R

The domination of gas turbines in most aero applications is obvious and marine use is considerable, but automotive propulsion has escaped so far. This leads to the question: Is this the end?

In retirement, I attend the Rolls-Royce Heritage Trust lectures in Coventry, which are often of interest. Sometimes the subject is of tremendous importance to me and one such occasion was when Steve Chilvers, Product Director of Marine Engines, visited and spoke of the WR21 engine propulsion system. This is for an engine of about 30,000 hp to drive passenger liners as well as naval craft to be available to marine operators, ship builders, etc from 2002. Hearing topics like heat exchangers, variable area nozzles, low emission combustion, world beating fuel consumption at low loads as well as full power, being described, you will understand how my interest was quickened. The WR21 is what I call a triple-shaft concept, incorporating an intercooler, a recuperator, variable area power turbine nozzles, and a means of removing all carbon deposit (shellac) from the gas side of the heat exchanger. The combustion system has nine chambers but covers a huge range of air/fuel ratios, occasionally cleaning the shellac off the gas-side matrix by running the compressor delivery air directly to the combustion chamber and causing the 'dirty' matrix to be at full gas-side temperatures – which should clear the matrix. It is also very carefully regulated in the air and fuel feeds to produce ultra-low emission characteristics at all times. Many of the turbine components come from the RB211/Trent class of modern generation engine, which augurs well for reliability and availability.

Having heard the talk and discussed the engine with the speaker, it seemed clear to me that Rolls-Royce is going into this most thoroughly and is aiming at all ships of significant size, where the steam turbine has dominated. Smaller naval vessels have used the benefit of rapid start and application of gas turbine power already, but the fuel consumption has been a penalty. The boot could now be changing feet, as the WR21 is expected to operate at a lower fuel consumption than the latest steam turbines and also the ship should gain some 15% space for more cargo or passengers – which equals revenue. No doubt the cost parameters will play a major role in the wide acceptance of the WR21 turbine, but my point in referring to this unit now is that this is definitely not the end in the marine field – maybe it will increase slowly, eventually to dominate here too. The variable power turbine nozzles are to maintain high cycle gas temperatures and are critical to the excellent, flat fuel consumption characteristic expected.

The WR21 is targeted to operate below 0.40 lb/hp/hr from 20% to 100% power – a remarkable achievement if it comes to pass. This synopsis is very scant and anyone could ask Rolls-Royce for more of their published reports if they are interested. I am personally impressed and optimistic.

The second group of people who won't give up turbines is nearer our own area of interest. Volvo have been working in small gas turbines since 1974 through a subsidiary, United Turbine (later Volvo Aero Turbine) in Sweden, and have run one or two cars since then. They did not see a near prospect for automotive engines at that time but they kept the pot boiling by embarking on an ambitious programme of hybrids. These combine two interacting power plants on one vehicle in a way that maximises the benefits of each, in a situation where the extra cost may be acceptable. Vehicle emission regulations are becoming tighter round the world and California has threatened that nil-emission zones will probably be demanded around crowded conurbations. The direction of Volvo's work is that a battery-powered electric motor would be the clean power source, recharged and boosted by a small turbine-powered generator. By 1995, Volvo had produced demonstration models of car, bus and truck powered by a series of hybrid units. In these vehicles, the wheels are powered by electric motors and the energy comes from either batteries on their own, the turbine-powered generator alone, or by both together. These demonstration vehicles have been around the world to note reactions, Volvo claiming very low emission levels with the relatively small turbine running and a reasonable mileage on battery alone.

One more arrangement strikes one as possible, though not yet seen, and that is to retain the hybrid concept and further reduce emissions by having a fuel cell to produce the electricity. If this is practical and economically viable, there may still be the need for a turbine to produce peak power – ie 'we' may still live! So take heart you turbineers, the world may still need our toys!

APPENDIX A
Schedule of Ministry of Aircraft Production contracts and orders received in connection with the development and production of Whittle superchargers

CONTRACTS ISSUED TO THE ROVER COMPANY LTD (PARENT)

Contract No	Date of Contract	
B.124307/40/C.28(b)	4 August 1940	Item 1 (a) Constructing and supplying four Whittle superchargers type W2B designed to run at a speed of 16,500rpm. Item 2 (b) Constructing and supplying two Whittle superchargers type W2 designed to run at a speed of 14,500rpm.
B.150845/40/C.28(b)	1 September 1940	(a) Investigations into the methods and equipment for the production of Whittle superchargers. (b) Investigations into the design of the first development engines of types W2 and W2B.
B.156333/40/C.28(b)	25 September 1940	Constructing and supplying two complete sets of spare parts for Whittle supercharger W2B.

The aforementioned contracts were terminated as from 28 February 1941 and the work continued on a 'Shadow' basis under the following order numbers: (marked *)

Order No	Date of Order	
*C/Eng/1271/C.28(b)	13 November 1941 and operating from 1 March 1941	(a) Investigations into the methods and equipment for the production of Whittle superchargers. (b) Investigations into the design of the first development superchargers of types W2 and W2B and subsequent superchargers of W2B type.
*C/Eng/523/C.28(b)	5 March 1941	Item 1 Constructing and supplying in accordance with the requirements of the Controller of Research and Development (CRD) 50 Whittle type W2B superchargers suitable in all respects for installation in the Gloster F9/40 aeroplane, or other aeroplanes as required by the CRD.

continued

*C/Eng/523/C.28(b) (continued)	5 March 1941	**Item 2** Constructing and supplying in accordance with the requirements of the CRD one Whittle type W2 supercharger. **Item 2A** Constructing and supplying in accordance with the requirements of the CRD one Whittle type W2 supercharger (originally ordered under Contract B.124307/40/C.28(b) and referred to in the Dept's letter 14/4/41. **Item 3** Constructing and supplying such modification parts as may be required and agreed by the CRD to be necessary to bring the performance and reliability of the superchargers ordered under Item 1 above up to the latest standard and specification requirements, assembling these parts. (a) Retrospectively in superchargers already built, if considered necessary, and (b) in new superchargers. Re-testing the superchargers after modification. **Item 4** Constructing and supplying such extra parts as may be required and approved by the CRD for development work at other than the works of or managed by the Rover Co Ltd (NB: details of parts supplied and consignees are to be furnished to the Minister). **Item 5** Constructing and supplying in accordance with the requirements of the CRD a mock-up supercharger and such parts as may be required and agreed by CRD to be necessary to bring this mock-up up-to-date from time to time.
C/Eng/1278/C.28(b)	14 November 1941	**Item 1** Constructing and supplying 550 Whittle type W2B superchargers or approved modifications thereof, suitable in all respects for installation in the Gloster F9/40 aeroplane, in accordance with specification (to be forwarded later).

C/Eng/1278/C.28(b) (continued)	14 November 1941	Item 2 Constructing and supplying such modification parts as may be required and agreed by the Controller of Research and Development and the Director General of Equipment to be necessary as an addition to the specification; assembling these parts retrospectively into superchargers constructed in accordance with the specification in order to bring their performance and reliability up to the agreed standard before delivery.
C/Eng.1581/C.28(b)	18 March 1942	Item 1 Supplying 12 packing cases for Whittle superchargers. Item 2 Supplying 18 transit stands suitable for Whittle superchargers. Item 3 Supplying to Drawing XAM81, ramps for use with the packing cases and transit stands supplied under items 1 and 2.
C/Eng/2545/C.28(b)	5 March 1943	Item 1 Carrying out in accordance with the requirements of and to a programme approved by the Controller of Research and Development (CRD) bench development testing of a duration not exceeding 500 hours on W2B superchargers, including the supply and fitting, or fitting only, of replacement or special parts, and stripping, examining and rebuilding the superchargers as required by the programme of testing, in accordance with specified inspection reports. (See Clause 6 of MAP Form 2010, revised April 1942).

APPENDIX B

EXPERIMENTAL DEPARTMENT REPORT.

Reference :- Lov/FRB. 2/JF. 2/3/45.

Subject of Report.

The possibilities of the gas turbine as a motive power for the automobile.

Summary.

A complete project is worked out for a gas turbine car engine of about 90 B.H.P. A preliminary scheme is given and the performance has been calculated for the engine fitted into 4¼ Litre Bentley. The performance figures show that for steady running on a level road fuel consumption varies from 25 miles per gallon between 20 and 50 M.P.H. to 17 M.P.G. at 80 M.P.H.

The turbine car shows many advantages when compared with the piston engined car, for instance the turbine unit should be cheaper, weigh less, have simpler auxiliaries, and require either no change gears or possibly a simple 2-speed gear and reverse.

With some development the gas turbine should be at least as reliable as the piston engine and probably as quiet or, even quieter. An outstanding advantage will be its smoothness.

Conclusions and Recommendations.

It is shown that the gas turbine is likely to be superior in all respects to the piston engine as a motive power for road vehicles. It is suggested that a turbine unit on the lines given in this report should be immediately put in hand if we are to enter the car field in the post war era and successfully compete with other firms who will undoubtedly follow this line of development.

Lov/FRB.

Full Report to:- Hs.   SGH.   E. Dor.   Rm. By. Da.
                 Lr.   Lov.   Rbr.Gr.   LH. SGH/RRJ.
                 SGH/LGD.   E/AAL.)   E/TGD. Lov/GLW.
                 Lov/JRJ.   SGH/CDD.   Rm/Slt.

Cover Sheet to:- Sg.   Rg.

FIG. 1.

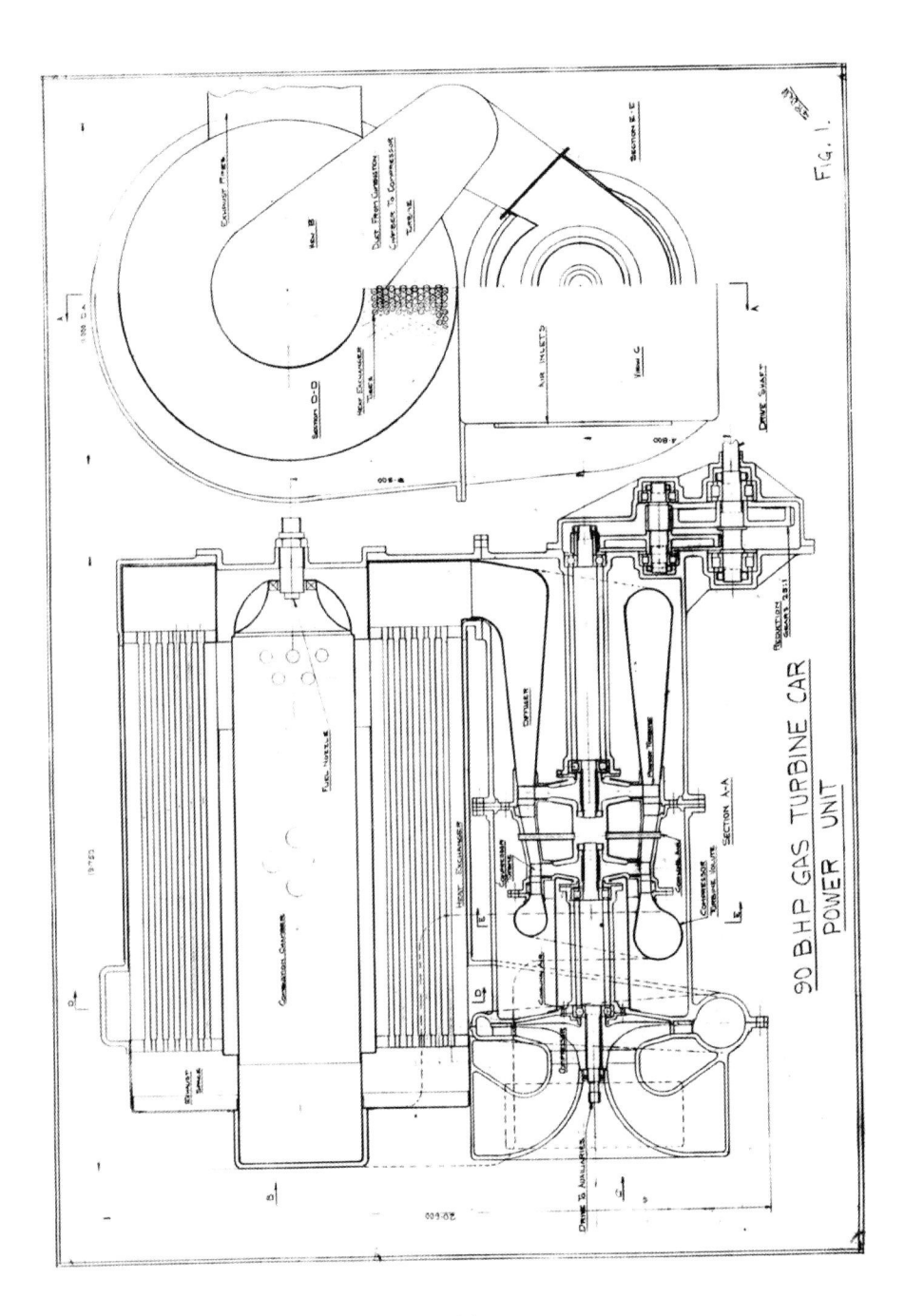

90 B.H.P. GAS TURBINE CAR
POWER UNIT

APPENDIX C

The Royal Automobile Club
83, Pall Mall, London, S.W.1.
ASSOCIATE SECTION

Please Address all Communications to the
GENERAL MANAGER
Quoting the following Reference

Telegrams: RACLUBIAN, AUTO, LONDON.
Telephone: WHITEHALL 4343

ET

REPORT OF TEST

ON

A ROVER GAS TURBINE CAR.

WEDNESDAY, 8th MARCH, 1950.

Held under the observation of the Engineering and Technical Department of the Royal Automobile Club.

The Rover Company, Ltd., Solihull, Birmingham, England, entered a Car fitted with a Gas Turbine Power Plant for Test.

The following leading particulars apply to this car :-

Wheelbase:	9' - 3".
Track:	4' - 4".
Type of Body:	Open.
Number of Seats:	2/3.

The general design of the car, apart from the power unit and transmission was on conventional lines, and its external appearance normal.

The Test was held on the Motor Industry Research Association Proving Ground circuit at Nuneaton, Warwickshire, and at the time of the Test the surface was dry, the weather calm, and the wind speed 5 m.p.h., temperature 54 F., Barometer 30.1".

The Entrant intended the Test to show the progress achieved by The Rover Co. Ltd., in the application of the Gas Turbine as a power plant for Road Vehicles.

Cont.

The Royal Automobile Club
83, Pall Mall, London, S.W.1.
ASSOCIATE SECTION

Please Address all Communications to the
GENERAL MANAGER
Quoting the following Reference

Telegrams : RACLUBIAN, AUTO, LONDON.
Telephone : WHITEHALL 4343

ET

- 2 -

The power plant which is mounted immediately ahead of the rear axle consists of a centrifugal compressor, with dual combustion chambers, having a single-stage compressor turbine, and an independent power turbine. The latter is positively coupled through gearing, incorporating a reverse gear, and drives a conventional rear axle.

The fuel used was kerosene.

The method of starting is by a normal car type electric starter, with push button on the instrument panel.

The time taken to start the power unit and to run up to idling speed was 13-1/5 seconds, and the car moved forward in a further 3-2/5 seconds.

Control is simple from standstill to maximum speed, and is solely by means of the accelerator pedal, the only other driving control being the brake pedal.

The reverse gear is operated by a lever which functioned normally.

The car was driven for five laps on a circuit of the Proving Ground measuring approximately 2.75 miles per lap, during which The Royal Automobile Club Observers travelled alternately as drivers and passengers.

No attempt was made to attain maximum speed, but during the course of the Test a speed exceeding 85 m.p.h. was readily attained, at which speed the compressor - turbine revolution counter indicated 35,000 r.p.m.

In a test of acceleration from a standstill, the car smoothly attained 60 m.p.h. in 14 seconds.

Although no provision for silencing the exhaust was observable, the volume of noise was not excessive or unpleasant, but was naturally accentuated during acceleration.

CHIEF ENGINEER,
ROYAL AUTOMOBILE CLUB,
ENGINEERING & TECHNICAL DEPARTMENT.

MEMBER OF,
ENGINEERING & TECHNICAL COMITTEE,
ROYAL AUTOMOBILE CLUB.

APPENDIX D

Comparison of compressor characteristics

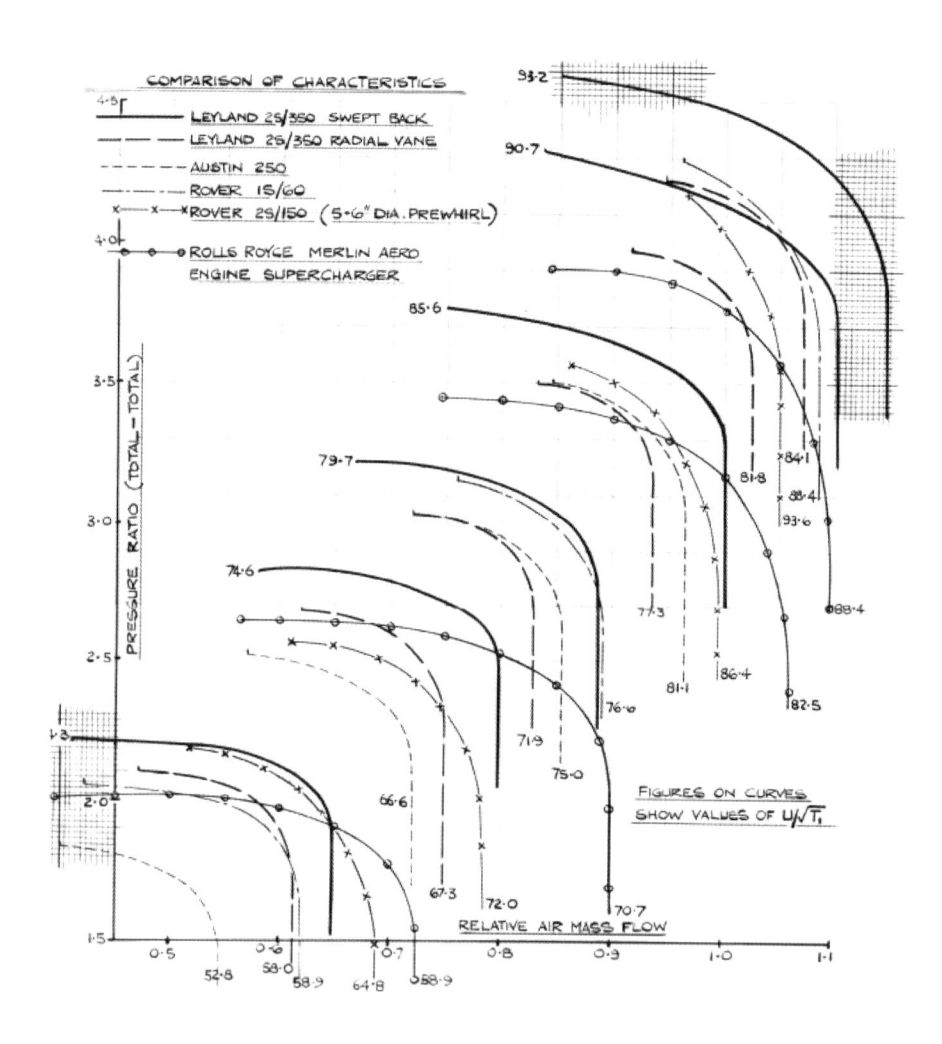

APPENDIX E
Summary of major Rover engine types

Start date	Engine	BHP	SFC lb/hp hr	Turbine inlet temp °C	Gas generator rpm	Air mass flow lb/sec	Pressure ratio	Engine weight lb
1948	T5	100	-	-	70,000	-	-	-
1950	T8	250	1.0	870°	40,000	-	-	-
1952	T6	100	-	-	-	-	-	-
1954	1S/60	60	1.5	800°	46,000	1.35	3.0	140
1956	1S/90	90	1.4	870°	46,000	1.95	3.0	140
1956	2S/100	110	-	870°	52,000	2.0	3.8	-
1959	2S/140	140	0.60	930°	65,000	2.0	3.9	480
1963	2S/150	150	0.9	930°	65,000	2.1	3.9	200
1965	2S/150R	145	0.55	950°	64,000	2.0	3.7	450
1968	2S/350R	370	0.40	1000°	38,200	3.75	4.0	980
1970	2S/350R Mk 2	400	0.43	1050°	38,200	4.1	4.5	-
1964	2S/75	96	1.1	930°	77,000	1.37	3.8	80
1966	TJ125	114lb thrust	-	930°	65,000	2.0	3.9	43

APPENDIX F
The Chrysler turbine car
Sandy Skinner

Exotic European and US turbine demonstration cars brightened the motor show circuit of the Fifties. In the real world only two motor manufacturers took their projects seriously as potential production machines – Rover and Chrysler.

Chrysler was unique among the great US car makers of the time in having a strong engineering ethos. The effective and brutally strong Hemi V-8 was introduced at a time when despite Merlin experience Ford and even the patrician Packard could only offer side valves. A later and large V8 was good enough to appeal to Bristol, Jensen and Facel Vega. The push-button Torqueflite automatic transmission was the best of the period, and torsion bar front suspension gave excellent handling.

Although not among the US manufacturers commissioned during WW2 to follow up Whittle developments, Chrysler made useful contributions to aero engine development and production. A Chicago satellite factory built more than 18,000 Wright 3350 engines for the B29 and the corporation designed the 36.4 litre, 2,500 bhp IV-220 inverted V-16 engine which flew in an experimental Thunderbolt. The unusual cylinder configuration of two V8s back-to-back presumably drew on automotive experience and some versions at least were built with axial superchargers, which must have helped entry into the turbine field.

Vehicle turbine development

Chrysler Laboratories started investigation of the gas turbine in the late 1930s under the direction of George Huebner, an engineer who was also an enthusiast and a first class communicator. Entry into the post-war gas turbine field was via a 1945 US Navy development contract for an economical turboprop. The decision to build automotive turbines was taken shortly afterwards and by the early 1950s experimental units were on test. In 1954 the Corporation demonstrated a 100 hp turbine in a Plymouth coupe and by 1956 the engine was sufficiently well developed to complete a four-day, 3,000-mile coast-to-coast demonstration trip. A second generation turbine announced in 1959 had significant improvements in design and materials. By this time the team included Giovanni Savonuzzi, ex-Fiat and Cisitalia.

At first sight the Chrysler turbine resembles the Rover unit. The basic architecture of the two engine families was very similar, with a single combustion chamber and a neat twin shaft compressor and power turbine arrangement. The most significant difference is that while Rover used a small inward flow radial compressor turbine Chrysler adopted the axial type. The low inertia of the gas generator section, due primarily to the use of an axial turbine, contributed to quick throttle response.

Turbine wheels with blade finish as-cast and minimal machining to maintain clearances were a good production proposition. The compressor wheel was the usual high strength light alloy.

A Huebner-directed metallurgical research programme developed low cost alloys with good high temperature performance to replace the high-cobalt alloys used initially for turbine wheels and other components. Moving to iron-base alloys reduced reliance on strategic metals at a time when the US Government regarded stockpiles as a strategic asset. It also reflected Chrysler's long-term interest in reduced production costs. In contrast, while Rover was aiming at design for low cost it accepted the use of materials with higher temperature capabilities in the interests of greater efficiency.

Two main families of alloys were developed. A 6% iron-aluminium alloy was used for most internal sheet metal and although failing to match the strength of more expensive austenitic Cr-Ni stainless steels, it showed much superior oxidation resistance at temperatures above 1,300°F (710°C).

Compressor wheel temperature at maximum power was 1,500°F (820°C) and under certain acceleration conditions could reach 1,825°F (1,000°C). This would normally call for lost wax cast 60+% cobalt alloys such as Vitallium, immortalised as a precision castable material by the great Sam Heron who won an argument by putting his cast Vitallium dentures on the table. The composition of the most widely used alloy, CRM-6D, is iron plus Cr 20%, Ni and Mn 5% each, and other alloying elements up to 1% each. Ultimate tensile strength plotted against temperature tracks Stellite 31 closely. A bonus is the lower specific gravity of the iron-based alloy, reducing inertia and gas generator acceleration time.

Huebner used a regenerator from the start, giving vastly better fuel consumption than non-regenerative automotive units. A single large rotating regenerator element on top of the turbine assembly on the earlier engines was replaced from the fourth generation unit onwards by dual units on either side rotating at 22 rpm. The 15.5 in (40 cm) cores were fabricated from one flat and one corrugated strip of 0.002 in (0.05 mm) stainless steel wound, furnace-brazed and then machined to provide a running surface. Graphite strips were used initially for the inner 'hot' seals but wore quickly and were replaced by another specially developed alloy, which is claimed to take on some of the characteristics of a ceramic at high temperatures. Regenerators made separate silencers unnecessary.

Engine accessories were straightforward. A single-stage reduction gear was coupled to a modified automatic transmission minus its torque converter. The single dynamo-starter on test cars proved mildly troublesome, as did the igniters: neither set of problems was regarded as serious. An automatic eight-second starting cycle with appropriate safety lockouts was initiated by turning the usual ignition key.

The public evaluation programme showed that engine maintenance was unlikely to be a problem: the commonest cause of complete stoppage was driver incompetence. Inevitable power loss in service was due to deposits, primarily traffic dirt – finely divided rubber plus oil, heat-cured on to the compressor diffuser passages. Soap and water or a small amount of ground pecan shells fed into the air inlet provided a cure.

The next generations: (1) engineering and evaluation

The third generation turbine (CR2A) appeared in a rather odd concept car at, among other events, the 1961 London Motor Show. Handouts stressed the benefits of a vast powered bubble top without making it clear whether there was a lockout to prevent it rising, complete with windscreen, while the car was moving.

The big engineering step forward was an automatic second-stage variable angle nozzle system immediately upstream of the power turbine. The angle of the 23 blades was changed continuously in response to throttle position by a hydraulic system powered by the main engine oil supply. This presented gas flow at the optimum angle of attack to the power turbine blades, varying effort progressively from full power, through medium and low speed economy positions, and finally to give engine braking.

The effect of the engine as a torque converter was enhanced and response to the throttle ('turbine lag') was substantially improved. According to Chrysler the first generation engine took seven seconds to accelerate from idle to full power, and the second three seconds. The variable nozzle cut this to one and a half seconds. The question of how much difference this made when driving is covered by the author in his detailed assessment.

The CR2A was installed in a pair of conservatively styled coupes, which were taken on an extended North American tour and demonstrated at Montlhery by US-based engineers. These were brave decisions since the engines were straight from the laboratory and no spares existed. The success of these trips encouraged an extraordinary public demonstration and evaluation exercise.

A fourth generation engine was immediately developed whose most obvious change was the use of twin regenerators. Internally, detail refinements reduced weight and bulk and further improved throttle response. This was installed in a run of purpose-built 2+2 coupes for public evaluation.

The US programme which followed is well documented. Fifty cars were built and loaned to a total of 203 private users from October 1963 to January 1966. Together they covered more than 1,100,000 miles with surprisingly little trouble, mainly confined to electrical accessories. The programme was closely supervised and documented from an engineering rather than PR point of view. Almost forty years later, the author still finds it impressive.

A variable turbine nozzle system seen here in a third generation unit; Chrysler turbines went a long way to solving the problem of turbine lag. Paired regenerators on the fourth generation unit helped economy.

The next generations: (2) but what's it like to drive?

Technical information in this paper is drawn mainly from published and unpublished Chrysler sources. This anecdotal assessment of the car on the road sets out to answer the subjective question: but what's it like to drive? The author might as well admit that he was either the perfect or worst possible tester, depending on your views, since by 1963 he had covered many more miles in R-R 20/25 and Frazer-Nash than in any car with IFS and automatic transmission.

European drivers didn't have an opportunity to try the standard-bodied cars in 1962. In 1963 Chrysler International took the fourth generation machine on a 23-city world tour, which included Paris and London.

The Paris event got off to a bad start with a malfunction while circling the Arc de Triomphe during the evening rush hour. After a '50 cent' solenoid had been replaced the hopelessly inadequate brakes were swapped for bigger and better ones from a special equipment car. Morale was not high.

During the Paris and London demonstration days, which followed, the car was 100% reliable.

The London event was based at Box Hill with a convenient dual carriageway leading to a filling station whose thatched roof over the pumps slightly surprised the US engineers. A frighteningly quick and almost wholly brakeless Chrysler 500J was hired from a US serviceman to give British journalists a taste of fast left-hand drive motoring before trying the turbine, and provided the author with entertaining personal transport for a week. He only spun it once.

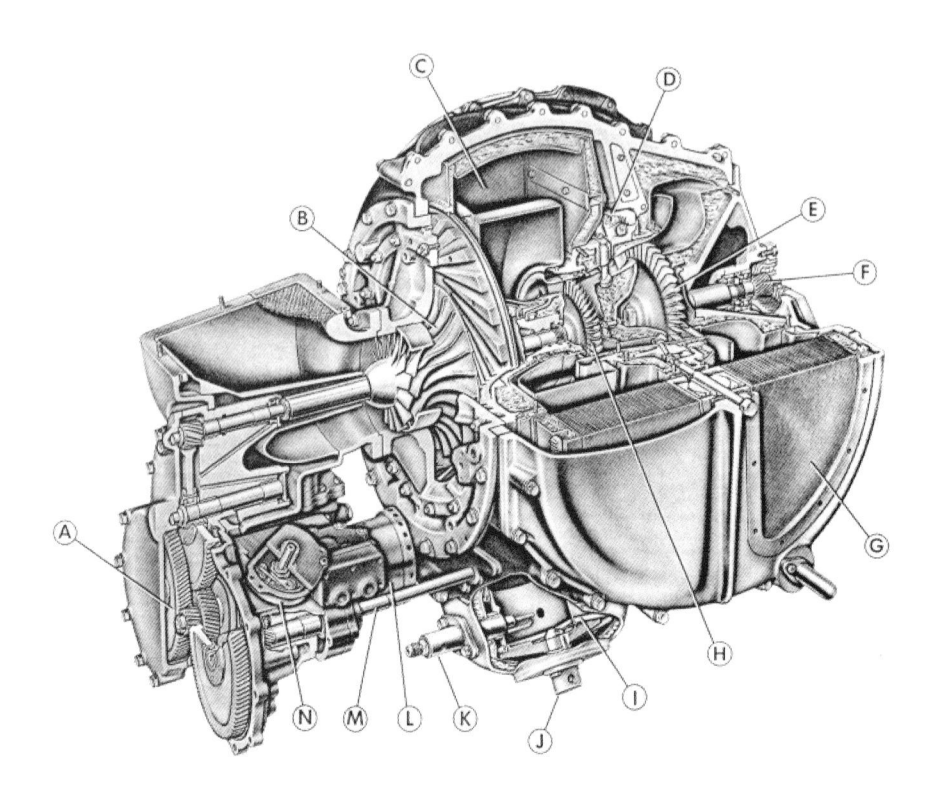

Main components of the twin-regenerator gas turbine: (A) accessory drive, (B) compressor, (C) right regenerator, (D) variable nozzle unit, (E) power turbine, (F) reduction gear, (G) left regenerator, (H) compressor turbine, (I) burner, (J) fuel nozzle, (K) igniter, (L) starter-generator, (M) regenerator drive shaft, (N) ignition unit.

Nominal performance of the 130 bhp turbine car was 115 mph top speed, almost certainly an over-estimate, and 0-60 in about 12 seconds if a hot start, running the engine up to speed against the brakes, was used. Fuel was either diesel or aviation kerosene, and the team was very cautious at the time about consumption figures. Later official figures were a creditable specific fuel consumption of 0.5 lb (0.23 kg) per bhp/hr, giving a claimed 19 mpg (15 litres/100 km) on motorways and 17 mpg (16.75 litres/100 km) in general US road use. Various figures have been given for stop-go consumption, all relatively poor: the author estimates that enjoyable town driving, making full use of acceleration, gave something like 10 mpg (28.5 litres/100 km).

Road manners were perfectly acceptable although steering was not up to Bugatti standards, and speed of response didn't quite match a chaingang Nash. Engine braking was about the same as a V8 with automatic transmission and the drum brakes were treated with respect. Noise level was not an issue: turbine whine was less than the blower whistle from the current generation of turbo diesel trucks.

As a generalisation, give and take performance with reasonably determined driving was rather better than the VDP-bodied 20/25 Royce up to 60 mph. At higher speeds, as one would expect, the turbine galloped away.

The celebrated turbine lag was not a problem in normal driving. Power shut off acceptably quickly when you lifted your foot, so there was no tendency to rush into corners. Hot starts were fun, and normal starts got the car away from the traffic lights while the usual dozy British driver of the time was still trying to find first gear. Published Rover turbine lag figures are significantly worse than for the nozzle-equipped Chrysler.

An interview with that distinguished engineer, the late Mr Tony Rudd, one of the very few people to have significant road experience of both the Chrysler and Rover turbines, added fascinating detail. He said that while commuting monthly on a large consultancy project for Chrysler he would *"Go to the airline desk to get a car park ticket and keys and take the freeway into Detroit. I always stayed in the same hotel and drove in to my first meeting the next morning. The car was sometimes a V8 and sometimes a Chrysler turbine. After a transatlantic trip on a 707 it didn't much matter"*. In other words, you got into the test series turbine and drove it on the public road like any other American car. He added, *"It didn't have sparkling acceleration, of course ..."* but agreed in telephone interviews and private correspondance that this could be down to the excessive weight of the complete package rather than the characteristics of the turbine engine.

Skinner enjoyed a considerable mileage in France and the UK in one of the Chrysler turbine world tour demonstration cars.

Mr Rudd was lukewarm about the Le Mans BRM-based Rover turbine cars, not on account of their performance but simply because they placed added loads on an already overworked Grand Prix racing organisation. Both went well: serial 00, the Hill/Ginther open 1963 car, won a special award and when radically rebuilt with a regenerative engine as the 1965 coupe came tenth on distance and was the first British car home in the hands of Hill and Stewart.

Forty years on, the author's view is that the demonstration Chrysler was entertaining personal transport in and around London. The Rovers presumably had vastly superior brakes and handling, but the Chrysler was unparalleled for what one might call social purposes.

The main Chrysler problem was weight. The steel body was built in Italy by Ghia and shipped to Chrysler's Detroit engineering laboratories for assembly. Although the engine unit weighed about half as much as a cast iron V8 of similar power the body was appallingly heavy, resulting in an all-up car weight of 4,100 lb (1,860 kg) – say, two tons on the road. This affected every aspect of performance.

Although hardly comparing like with like, the author would like to try the Formula One BRM-based Le Mans Rover coupe, weighing 1,960 lb (845 kg) wet with driver, fitted with a nozzle-equipped Chrysler engine. It could be a splendid sports car.

Envoi

The turbine car was a beautiful theory murdered by a set of brutal facts. The author greatly enjoyed playing with the Chrysler but would find it very hard to justify its development and 50-car build and test programme, whose true cost has never been revealed. There were advantages compared with the piston engine, but none were compelling enough to justify enormous spend on its replacement.

The world tour car and most of the US test vehicles were, in the Rolls-Royce phrase, reduced to produce. This wasn't vandalism: a few went to museums and every major manufacturer knows that selling experimental vehicles to the public stores up service problems in the future. Work limped on into the '70s with slightly improved engines funded mainly by the Environmental Protection Agency; when Federal funds dried up, so did the turbine programme. It was a gallant effort.

APPENDIX G
The Fiat Turbina
Andrew Nahum

On an April day in 1954, the famous banked test track on the roof at Fiat's Lingotto factory echoed to a new kind of sound. There was a high-pitched whine, rising to a jet plane scream. The noise was coming from what looked like a Buck Rogers rocket ship that had emerged from a secret experimental shop. It looked unlike any car the Torinese auto workers had ever seen. Fiat were out to show that despite the economic damage and technical stagnation of the war years, they were back in the running and in full command of the latest modern technology.

The Second World War had been a disastrous period for Fiat. The German and Allied armies had fought their way across the country, factories were bombed and the internal Italian dispute between the Fascists and the resistance organisations spilled over into the Fiat works. Fiat management was caught between two fires; the Germans wanted production of war materials; the Resistance demanded an end to Axis collaboration. For a while, Agnelli and some top managers were suspended by the Resistance who appointed their own technical committees to run the company.

In the immediate post-war years, Fiat set to repairing their own damage and the revival of the motor industry made a vital contribution to North Italian economic revival; it also played an important symbolic part in reviving Italian self esteem. One particular miracle had been the sight of seven tiny Cisitalia racing cars competing in Turin's Valentino Park in 1946. Pre-war, Italians had been used to seeing their competition cars take on the world, but creating a new racing car the year after the war had still seemed like a triumph in the face of adversity – even if the mechanical parts were derived from humble Fiat 1100 and 500 models. The Cisitalia had been designed by Dante Giacosa, moonlighting from Fiat in the last years of the war. He was to become famous as Fiat's long-term engineering boss and as life got back to normal he severed the Cisitalia link to design less frivolous cars for Italy's essential post-war transport. These included the tiny and highly utilitarian Fiat 600 and Nuova Cinquecento. In great secrecy, he also initiated the gas turbine car project, partly for research, but also to make a clear propaganda statement about Fiat technical prowess.

Giacosa had first glimpsed parts of the then revolutionary jet engine when it was a military secret during the war. The Germans had submitted a component for Fiat to manufacture. It was known to be part of an aero engine – but was obviously nothing to do with the traditional Daimler-Benz piston engines that they were making under sub-contract. Subsequently he realised it was the turbine for the Jumo jet engine that powered the Messerschmitt 262 fighter.

Fiat, of course, had its own aviation division, and in the Thirties had produced the fearsome 3,000 hp double engine (two V12s coupled together) that had raised the world air speed record to 709 km/hr in a Schneider Trophy seaplane. Post-war, they took out a licence to build the British de Havilland Ghost jet engine. The turbine car engine came to look a bit like a miniature Ghost but the design in fact was developed entirely within the car division; aviation turbine specialists were too scarce to spare for the off-beat experiment. It used tandem compressors like the Rolls-Royce Dart and the intake was reversed to shorten the engine. The auto engineers soon found that a small road transport unit posed problems that almost were more severe than contemporary full-size aircraft jets. Paradoxically, the smaller size of the turbines and combustion chambers meant they were subject to more severe heating, for proportionately the surface area of metal exposed to burning gas is greater.

While the engine was being developed, the 'Special Coachwork Section' developed a body shell for it. This side of the project was under the direction of a visionary engineer, Luigi Fabio Rapi. Rapi had an unusual combination of skills, having studied mechanical engineering, aircraft construction and architecture, and had worked for a while at Isotta Fraschini before joining Fiat. In addition to an understanding of structures and production methods, he could visualise dramatic body shapes and also had an intuitive feeling for aerodynamic efficiency. His first project at Fiat was the stylish 8V sports coupe, launched in 1952.

In England, Rover were content to try their gas turbine in a near standard saloon shell with the 'glasshouse' cut off to make a roadster. Fiat, though, wanted something special to highlight the technical capability they were displaying with the engine itself. Rapi borrowed the excellent fully independent suspension from the 8V (double wishbones all round with anti-roll bars), but he started with a clean sheet of paper from the body. The rear of the car was taken up with the bulky turbine unit, so 22 gallons of kerosene fuel were accommodated in two long tanks running inside the sills. Six lead-acid batteries were needed to provide 24V and enough current to spin up the gas turbine, and with these placed right in the nose of the car, the weight distribution was a near ideal split of 42/58 per cent front to rear.

It was, however, in the aerodynamic form of the car that Rapi excelled himself. The tailfins were not merely a sci-fi accessory – they were judged essential for stability at the very high speeds thought likely, while the general shape was slippery in the extreme. The drag coefficient of the full-size car was never determined but from wind-tunnel tests on 1/5th scale models it was calculated to have an ultra-low Cd in the region of 0.14.

During test bed engine runs, the Fiat engineers soon got the message that was also dawning in the minds of other gas turbine experimenters at GM, Chrysler and Rover. Without a heat exchanger, the gas turbine is a very inefficient powerplant for a road vehicle. Its operating cycle relies on passing a very large quantity of air (the gas turbine burns an ultra-lean mixture) and the result is that a large proportion of the precious heat energy is lost down the tailpipe.

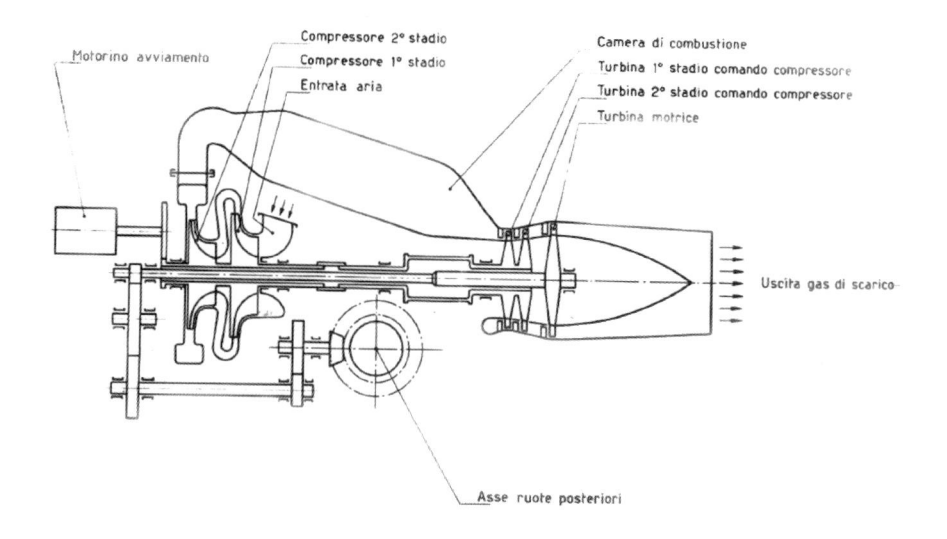

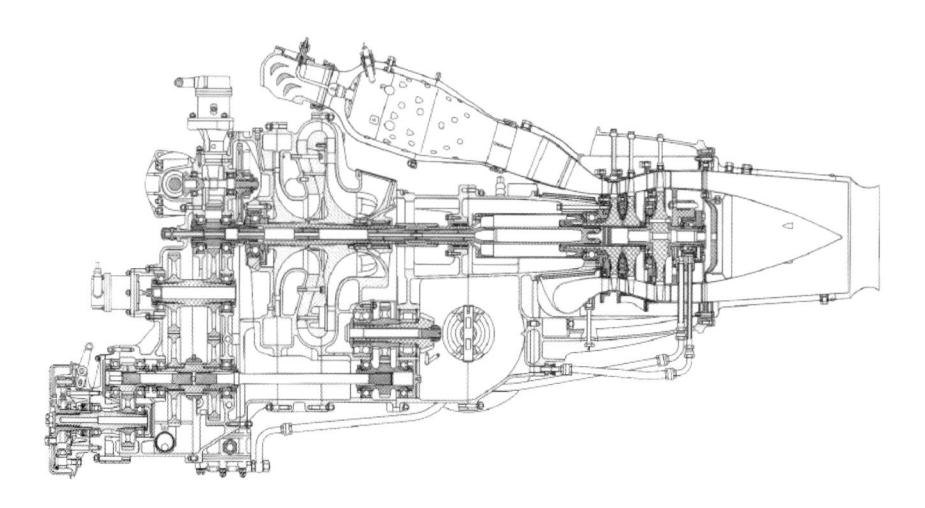

Sectional view of Turbina's engine. Although having only three combustion cans, it resembles de Havilland Ghost in the combustion chamber design, though it has tandem centrifugal compressors like the Rolls-Royce Dart. Intake is reversed to shorten the engine. *(Fiat)*

The futuristic Turbina body design (Fiat)

Heat exchange

The answer is an efficient heat exchanger, which can draw waste heat from the exhaust gas and use it to pre-heat the incoming air. The Rover company believed they had one when they were on the verge of putting their gas turbine into production in a top of the range car. Doubts developed about its durability and the release of the model was shelved. Since then, an effective heat exchanger has so far remained an unobtainable Philosopher's Stone for advocates of the gas turbine for road transport. Thus it became clear that the Fiat gas turbine was not even a long-range prototype for a road vehicle engine. (Not least among its problems would have been the large volume of exhaust air at 400°C swirling around the traffic and melting the stockings on pedestrians at crossings!)

Though the practicality of the power unit was in doubt, the original aim to demonstrate Fiat's 'hi-tech' capability was still valid and as completion of the actual car was imminent, it was decided to press ahead with the demonstration. The driver was to be veteran test driver, Carlo Salamano, who had joined Fiat in 1911 and during his long reign had stamped his mark on every new model the company released. He took the Turbina round the Lingotto roof top track, joking that he had no idea how fast it would go, and that he should, maybe, wear a parachute in case it flew off the banking.

A few days later, the car was taken to Turin's Caselle airport for tests, again driven by Salamano. It must have been far from easy to drive, for there was no form of automatic control fitted of the type that is now commonplace on gas turbines. The accelerator was coupled directly to a throttle valve and, as a result, under acceleration, Salamano had to monitor a bank of instruments giving turbine temperature (to avoid overheating), and turbine rpm (to avoid a destructive overspeed), as well as keeping an eye on the secondary 'power turbine' rpm and the road speed.

The turbines were designed to run at more than 30,000 rpm, but for the Caselle tests they were held to 27,000, at which speed the engine gave about 150 hp. At this power level, the road speed was said to be in excess of 120 mph. After only two days of trials, the car was demonstrated to the press and then exhibited at the 1954 Turin motor show where its futuristic lines attracted enormous interest. Its only other outings were for some brief testing at Monza and a few demonstration laps before the Rome Grand Prix, at the Castel Fusano circuit.

One slight sadness in the story is that we never learned what the true speed potential of the car was. 120 mph seems a very conservative figure though the engine was throttled to 150 hp. At 2,200 lb, the vehicle is not particularly light but with a Cd of 0.14 its aerodynamic penetration was superb, even by experimental standards today. In a 1958 Fiat technical paper, the design output was said to be in the region of 300 hp and it does not seem fanciful to suggest that with full horsepower unleashed, the car would have well exceeded 200 mph – if the body did not display too much lift (contemporary aerodynamics could devise shapes that would penetrate the air; stability was a different question). Though fearless, Carlo Salamano was by then getting on in years and no one else was entrusted with it. After all, there was no possibility of the car or engine being developed further and, therefore, no sensible reason to risk driver or machine in an attempt to discover its ultimate performance capability.

Fiat's test driver, left, with cap Salamano, with the Turbina at Caselle airport   *(Fiat)*

Today, the Turbina is displayed in Turin's famous Carlo Biscaretti motor museum, although it is also on show from time to time in Fiat's own Centro Storico. It is a splendid machine and, for students of motoring originality who find themselves in Northern Italy, well worth a pilgrimage.

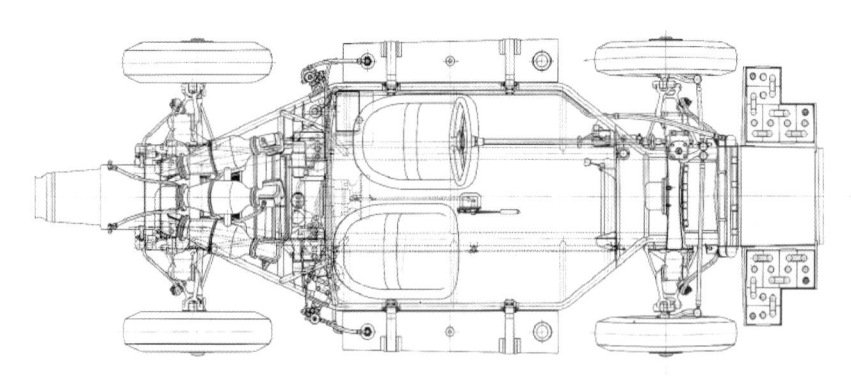

The neat installation of the engine in the Turbina chassis as illustrated in drawing and photograph

The Historical Series is published as a joint initiative by the Rolls-Royce Heritage Trust and The Rolls-Royce Enthusiasts' Club.

Also published in the series:

No 1	*Rolls-Royce - the formative years 1906-1939*
	Alec Harvey-Bailey, RRHT 2nd edition 1983
No 2	*The Merlin in perspective - the combat years*
	Alec Harvey-Bailey, RRHT 4th edition 1995
No 3	*Rolls-Royce - the pursuit of excellence*
	Alec Harvey-Bailey and Mike Evans, SHRMF 1984
No 4	*In the beginning – the Manchester origins of Rolls-Royce*
	Mike Evans, RRHT 1984
No 5	*Rolls-Royce – the Derby Bentleys*
	Alec Harvey-Bailey, SHRMF 1985
No 6	*The early days of Rolls-Royce - and the Montagu family*
	Lord Montagu of Beaulieu, RRHT 1986
No 7	*Rolls-Royce – Hives, the quiet tiger*
	Alec Harvey-Bailey, SHRMF 1985
No 8	*Rolls-Royce – Twenty to Wraith*
	Alec Harvey-Bailey, SHRMF 1986
No 9	*Rolls-Royce and the Mustang*
	David Birch, RRHT 1997
No 10	*From Gipsy to Gem with diversions, 1926-1986*
	Peter Stokes, RRHT 1987
No 11	*Armstrong Siddeley - the Parkside story, 1896-1939*
	Ray Cook, RRHT 1989
No 12	*Henry Royce – mechanic*
	Donald Bastow, RRHT 1989
No 14	*Rolls-Royce - the sons of Martha*
	Alec Harvey-Bailey, SHRMF 1989
No 15	*Olympus - the first forty years*
	Alan Baxter, RRHT 1990
No 16	*Rolls-Royce piston aero engines - a designer remembers*
	A A Rubbra, RRHT 1990
No 17	*Charlie Rolls – pioneer aviator*
	Gordon Bruce, RRHT 1990
No 18	*The Rolls-Royce Dart - pioneering turboprop*
	Roy Heathcote, RRHT 1992
No 19	*The Merlin 100 series - the ultimate military development*
	Alec Harvey-Bailey and Dave Piggott, RRHT 1993

No 20	*Rolls-Royce – Hives' turbulent barons* Alec Harvey-Bailey, SHRMF 1992
No 21	*The Rolls-Royce Crecy* Nahum, Foster-Pegg, Birch RRHT 1994
No 22	*Vikings at Waterloo - the wartime work on the Whittle jet engine by the Rover Company* David S Brooks, RRHT 1997
No 23	*Rolls-Royce - the first cars from Crewe* K E Lea, RRHT 1997
No 24	*The Rolls-Royce Tyne* L Haworth, RRHT 1998
No 25	*A View of Ansty* D E Williams, RRHT 1998
No 26	*Fedden – the life of Sir Roy Fedden* Bill Gunston OBE FRAeS, RRHT 1998
No 27	*Lord Northcliffe – and the early years of Rolls-Royce* Hugh Driver, RREC 1998
No 28	*Boxkite to Jet – the remarkable career of Frank B Halford* Douglas R Taylor, RRHT 1999
No 29	*Rolls-Royce on the front line – the life and times of a Service Engineer* Tony Henniker, RRHT 2000
No 30	*The Rolls-Royce Tay engine and the BAC One-Eleven* Ken Goddard, RRHT 2001
No 31	*An account of partnership – industry, government and the aero engine* G P Bulman, RRHT 2002
No 32	*The bombing of Rolls-Royce at Derby in two World Wars – with diversions* Kirk, Felix and Bartnik, RRHT 2002
No 33	*Early Russian jet engines - the Nene and Derwent in the Soviet Union, and the evolution of the VK-1* Vladimir Kotelnikov and Tony Buttler, RRHT 2003.
Special	*Sectioned drawings of piston aero engines* L Jones, RRHT 1995
Monograph	*Rolls-Royce Armaments* D Birch, RRHT 2000

The Technical Series is published by the Rolls-Royce Heritage Trust.

Also published in the series:

No 1	*Rolls-Royce and the Rateau Patents* H Pearson, RRHT 1989
No 2	*The vital spark! The development of aero engine sparking plugs* K Gough, RRHT 1991
No 3	*The performance of a supercharged aero engine* S Hooker, H Reed and A Yarker, RRHT 1997
No 4	*Flow matching of the stages of axial compressors* Geoffrey Wilde OBE, RRHT 1999
No 5	*Fast jets – the history of reheat development at Derby* Cyril Elliott, RRHT 2001
No 6	*Royce and the vibration damper* T C Clarke, RRHT 2003

Books are available from:
Rolls-Royce Heritage Trust, Rolls-Royce plc, Moor Lane, PO Box 31, Derby DE24 8BJ

ROLLS-ROYCE
HERITAGE TRUST